Apollo EECOM:

Journey of a Lifetime

All rights reserved under article two of the Berne Copyright Convention (1971).
We acknowledge the financial support of the Government of Canada through the Book
Publishing Industry Development Program
for our publishing activities.

Published by Apogee Books an imprint of
Collector's Guide Publishing Inc., Box 62034, Burlington, Ontario, Canada, L7R 4K2,
http://www.cgpublishing.com
Printed and bound in Canada

APOLLO EECOM: Journey of a Lifetime by Sy Liebergot (2nd Edition)
©2006 Apogee Books & Sy Liebergot

Apollo EECOM:
Journey of a Lifetime

by
Sy Liebergot

With David M. Harland

An Apogee Books Publication

This book is dedicated to all those people who made a difference in my life. Sometimes the lessons they provided were painful and difficult to accept; other times the lessons brought clarity and joy. Gratefully, I learned sufficiently to deserve my life partner and wife, Craig, who brought Bogie and Sam into my life.

- Contents -

Acknowledgments		7
Forewords by Ron Howard and Clint Howard		9
Introduction		10
Mission Control Center		11
EECOM		12
Prologue: 1936 World Events		13
Part One: The Early Years and Decisions		
1.	April 13, 1970: I Was Thirty-Four Years Old…	16
2.	March 12, 1942: I Was Six Years Old…	17
3.	February 15, 1936: Born	18
4.	1939: The Journey Begins	21
5.	1941: Philadelphia Memories	23
6.	1942: Kingston, New York	25
7.	Elementary School: First Grade	26
8.	World War II: Signs of the Times	27
9.	1942: The Destruction of Ida Soloff	28
10.	Foster Homes	30
11.	1943: Together, Again	32
12.	1944: Kingston, Part II	33
13.	Second Grade: A Lasting Impact	34
14.	Stability … At Last?	35
15.	Babies, Beatings, Pasta	37
16.	1945-1950: Signs of the Times	39
17.	Growing Pains	40
18.	Kingston High School: Music Appreciation	42
19.	December, 1950: Back to Philadelphia - A New Culture	43
20.	December, 1950: Northeast High School	46
21.	Classmates	49
22.	High School Years: Signs of the Times	51
23.	1953: The Philadelphia Inquirer	52
24.	1954: Decision Time	54
25.	September, 1954: Army Enlistment	55
26.	Life After Sol	56
27.	1954-1957: The Army Years - Basic Training	59
28.	1955: Odyssey of Unexpected Destinations	62
29.	Fort Huachuca: The Beginning of Hell	63
30.	No, This Is Hell - Yuma Test Station	65
31.	A New Life: Los Angeles, California	74
32.	Family and College	76
Part Two: Mission Control		
33.	Shaky Start	81
34.	July, 1964: Houston, Texas	83
35.	Flight Controls Beginning	86
36.	The Singin' Wheel	87
37.	Flight Control: The New Mistress	90

38.	Mission Control: Origin and Fate	92
39.	A Little Tutoring Here: Mission Operations Control Room	95
40.	Essential Equipment: The Flight Controller Console	98
41.	The Flight Controller: Apollo Era	101
42.	Embracing a Tradition: Becoming an EECOM	116
43.	Apollo: The Early Missions	122
44.	Apollo Continued: The Lunar Missions	131
45.	Apollo 13: The Longest Hour	138
46.	Apollo 13: Trail to the Movie	152
47.	Apollo 14: Bruce McCandless	157
48.	Apollo 15: Some Flight Controller Examples	160
49.	Apollo 16: Bodily Function	166
50.	Skylab: MSFC Steps to the Fore	167
51.	Astp: The Russians are Coming…	175

Part Three: Life After Mission Control

52.	EECOM No More: Set Adrift	184
53.	1972: Beginning Anew … Several Times	186
54.	Finally Getting It Right	191
55.	Travels and Food	193
56.	Life After NASA	196

Epilogue: Reflections — 199

Appendices

A.	The Mission Control Patch	201
B.	Apollo 13: O2 Tanks Chronological History	203
C.	Apollo 13: EECOM Console Displays	208
D.	Apollo 13: Number Coincidences	212
E.	Those Who Have Passed: In Memorium	213
F.	Acronyms and Abbreviations	214

— Acknowledgments —

Five years ago my wife Craig suggested that I should write my autobiography. I thought that this was a great idea, but how does one begin such a project? I had no answer, so I did nothing. However, when Tom Hanks found an opportunity to echo Craig's sentiments at an *Apollo 13* fundraising affair, I began an effort in that direction.

Over the years I have read a number of books that were written about our nation's space program, and most recently books by Chris Kraft and Gene Kranz relating their lives and Mission Control experiences. Although these were fascinating, I was left with the feeling that the story of the Apollo Flight Controllers remained untold. So finally, early in 2002, I wrote out a chronological outline of my life to serve as the core of my book. Michael Lennick, a Discovery Channel producer and new friend, spurred me on with encouragement and advice. He recommended that I send an outline to Rob Godwin of Apogee Books who, to my delight, told me: "You have a book, keep going. I'd like to publish it."

In seeking a technically-minded editor, I was introduced by W. David Woods, the creator and curator of the Apollo Flight Journals on the internet, to David M. Harland, a prolific author of books about space. A tireless and hard taskmaster, he kept my tenses consistent and the story flowing seamlessly, and he was able to correct occasional faulty memories of technical issues. David is a Brit living in Scotland, and it was humorous to both of us when cultural differences arose when I used U.S.A. slang or idioms. Although this is my story, told in my words, it would not be the book that it is without David's help.

W. David Woods, in his private role as a space historian, always found the time to research a mission event for me, and send me copies of the relevant documentation.

Bob Legler, Lunar Module Apollo Flight Controller and the keeper of the Mission Control statistical history of the Gemini, Apollo and Shuttle programs, provided accurate manning lists and helped me to recreate the camaraderie of the early times. Now retired, perhaps Bob will have the time to record all of his experiences and anecdotes in a book of his own.

Jim Hannigan, former Lunar Module Systems Branch Chief, drew my attention to the timely development of the LM Lifeboat Procedures that he authorized as early as *Apollo 10*, and I was delighted to find an opportunity to recognize the unsung efforts of the LM Flight Controllers.

Other Flight Controllers who answered my many questions about their personal experiences were: Ed Fendell, Gary Coen, Gerry Griffin, Jerry Bostick, Chuck Deiterich, Milt Heflin, Neil Hutchinson, Jay Greene, Hal Loden, Bob Heselmeyer, and Merlin Merritt and Gene Kranz.

Glen Swanson, formerly the NASA Historian at the Johnson Space Center, researched historical events for me and provided copies of documentation. Rebecca Wright, Program Manager of the Oral History Project, dubbed my 33 year-old console voice loop tapes onto CDs. I must thank Fred Schoeller for permission to reproduce images and text first published by Arts & Letters on the CD-ROM *Apollo 13: A Race Against Time*. Other official images were secured with the help of Mike Gentry in the Media Resource Center at JSC and David Sharron in the JSC Film Repository.

Unafraid to read and comment on my manuscript were friends Mark Caterina, Fred Schoeller and Ginger Hanks.

Importantly, I am grateful to my nephew, Dr. Stephen Pulley, for the opening thoughts in the Introduction. He helped me to realize "the essence of my life."

Finally, I must thank my sister Phyllis Pulley and brother Marvin Liebergot for their painful childhood recollections and photos.

— Foreword by Ron Howard —

My whole perception of the film *Apollo 13* shifted when I went on my first research trip to the Johnson Space Center and sat in the MOCR with a dozen of the flight controllers who had been involved in the mission, including Sy Liebergot. I had always understood the original screenplay as a heroic survival story of three brave astronauts. However, while listening to these men reflect on their role in the crisis, I began to realize not only could the movie accomplish that truth but also shed light on the heroics of the men of Mission Control. Not Rambo-like, acrobatic heroics of violence but instead, the heroics of the application of accrued knowledge, combined with a tremendous emotional personal drive. I had come upon a directorial opportunity to dramatize the power of the mind; a mind applied under great stress. It might not be leaping from building to building but what the flight controllers did was a Herculean accomplishment. It was a great test and those tests always make great drama. A key moment in the mission, dramatically speaking, was the moment Sy, portrayed by my brother Clint in the film, had when his recommendation from the EECOM console was to shut a fuel cell's reactant valves. The objective of the mission was landing on the Moon and Sy's call not only ended that, but also gave full dimension of the disaster they were facing. Aside from his involvement in *Apollo 13*, Sy's life experiences should be held as an example of achievement. Young people ought to read this book and take note of a life well done.

— *Ron Howard, Director of the Apollo 13 film*

— Foreword by Clint Howard —

I've had many perks in my forty-plus years in the entertainment business but few rise above the friendship that developed with Sy Liebergot. I remember the nervous anticipation I had prior to my first phone call with Sy. I was getting a chance to quiz the man I was to play as White EECOM in the film, *Apollo 13* and I didn't want to come off as a fool. As we talked, I remember Sy being very patient with me. I also remember him saying something that would be a huge help for me preparing for the role, "I was no steely eyed missile man, that's for sure." Sy was very humble, referring to himself as just an engineer and although he was a veteran of the manned space program, he recounted the lack of bravado he felt during the days of crisis on *Apollo 13*. After getting to know Sy over the years and now having read his memoirs, I realize what a remarkable man he is. I know he'll never say this about himself, but Sy Liebergot is a hero. Heroes rise above personal adversity to make great achievements. From where Sy came to what he's made of his life is a great achievement. NASA gave a young man, fresh out of college, an opportunity to be part of a new frontier and Sy didn't disappoint. To be part of a team that started with a blank sheet of paper and developed the hardware and procedures for manned space travel is pretty special.

— *Clint Howard, White EECOM, Apollo 13 film*

— Introduction —

This book is an autobiography of my journey of personal growth from uncertain beginnings to maturity and an experience-rich, fulfilled life.

In writing this book I was surprised to discover that I had suppressed some early memories and that I recalled so many with absolute clarity. It would have been far easier to leave them buried. The search for my history has allowed me to view my life in perspective and more importantly provided an impetus to reach out and attempt to reconnect with half-siblings Marvin, David and Denise.

My sister Phyllis and I were served up a terrible upbringing and the resultant aversion pushed me all the way to California away from family and onto a path that lead to my involvement in one of the greatest endeavors man has achieved so far. For Phyllis it had the opposite effect by pushing her into a tight-knit group of relatives and adopted relatives and friends while anchored to a small home in Roslyn, Pennsylvania.

> *The zenith of the human experience is when a man takes tragedy and is able to convert it into something positive and fulfilling. That is the essence of my life. I chose to rise above my predicament. Hopefully, my journey may serve as an assuring example.*

The story falls readily into three parts:

Part One deals with my early years of poor beginnings with great disadvantages, a situation precipitated by a father who was a lifelong gambler. His addiction destroyed any chance for a normal childhood and he took us all down with him. I have no recollection of being hugged or kissed by my parents; only of abuse and lack of affection. This part also reviews my life as I struggled with the decisions for the future: the military and college years, my first marriage and the beginning of my own family.

Part Two examines and depicts my career years as an engineer, reaching a personal apex as a NASA Flight Controller during the golden time of the Apollo Program. After all the years of giving interviews, speeches, and contributing to others' books and documentaries, I came to the realization that the real flight controller story has not yet been told. Most books written about our space program or Mission Control have either dealt mainly with the astronauts or were "top down" accounts, such as seen from the Flight Director position or the Program management.

This part of my story is told from "the bottom up"; it represents an insider's perspective of the early days, explaining what it was like to work for various Flight Directors, the flight controllers' reactions to flight situations, the flight controllers' interactions with the astronauts, the flight controllers' hangouts, the *esprit de corps*, the inside anecdotes and jokes, and *the absolute dedication of the flight controllers, whose whole existence was focused on mission success and astronaut safety.*

Part Three recounts my post flight controller years, the career changes and challenges involved in developing a space station, followed by building of the Space Station Extravehicular Activities (EVA) trainers for the Neutral Buoyancy Training Facility. There is a recounting of continuing new personal interests, friends, other marriages, and importantly, my personal growth that threads through this journey.

I leave these recollections as a legacy for my children, siblings, and the generations that follow.

Sy Liebergot, April 2003

— Mission Control Center —

The Space Task Group (STG) was formed at Langley Field, Virginia, in November 1958, with specific management responsibilities for the design and development of a spacecraft for project Mercury, selection and training of astronauts, and control of the flights. In 1962, the STG was relocated to the new Manned Spacecraft Center (MSC) that was being built in Houston, Texas following the successful completion of Project Mercury in May 1963; the Control Center at MSC assumed the responsibility for the Gemini and Apollo Programs.

During Gemini, Apollo, and now Shuttle flights, control is exercised from the Control Center from lift-off through recovery. The International Space Station has been added to the exercise of constant surveillance that is maintained over vital spacecraft systems. Based on the performance of the in-flight spacecraft as analyzed by a team of flight controllers, key decisions are made as necessary to accomplish flight objectives and assure mission success.

In March 1973, MSC was renamed the Johnson Space Center (JSC), in honor of Lyndon B. Johnson, the 36th President of the United States.

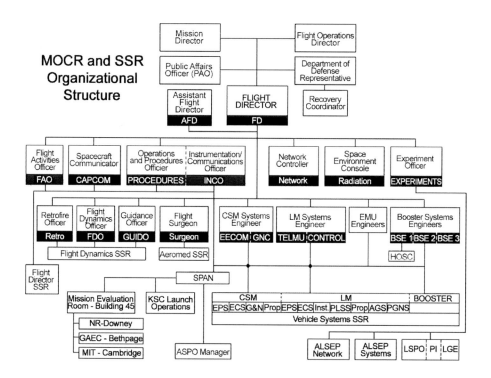

— EECOM —

EECOM is an acronym introduced for the Mercury Program. It originally stood for Electrical, Environmental and COMmunication systems. This historic radio call sign is unique to manned space flight operations.

The EECOM position in the Mission Operations Control Room (MOCR) in the Mission Control Center is steeped in tradition. Men who sat at the EECOM console during Apollo inherited standards of professional behavior and performance that gave newcomers to the position some pause. Ongoing Gemini flights and the addition of Apollo operations required training of additional EECOMs, subjecting them to a steep learning curve.

The Apollo EECOM Flight Controller was responsible for the life support systems of the Command and Service Module (CSM), which amounted to roughly half of the systems of that spacecraft. These were the systems that provided all electrical power and its distribution, heating and cooling, cabin atmosphere pressure control, breathing oxygen, cryogenic hydrogen and oxygen for the fuel cell electrical power plants, the sequential system that controlled the separation events and the parachutes, and many of the mechanical systems. The EECOM was also responsible for the CSM communications system through Apollo 10; thereafter that responsibility was moved to a new console named INCO. The EECOM position responsibilities became even further diluted on the Shuttle program.

The Flight Directors historically saw EECOM as a catch-all: if a function didn't fit anywhere, then it belonged to EECOM. An EECOM was willing to take responsibility for situations or anomalies not claimed by anyone else. "That's yours, isn't it EECOM?" was a common Flight Director query. Consequently, an EECOM was required to develop a broad understanding of the spacecraft systems and their operation.

— Prologue: 1936 World Events —

The world I was born into was sown with the seeds of World War II. Adolf Hitler's army had just occupied the Rhineland and he formed an "axis" with Italy's Benito Mussolini to join on the side of General Francisco Franco's military uprising in the Spanish Civil War.

The world of professional sports welcomed Joltin' Joe DiMaggio to the baseball major leagues who, as a rookie, batted .323 to help the New York Yankees win the American League pennant, and Jesse Owens, a black sharecropper's son from Alabama, won four gold medals in the Berlin Olympics showing up dictator Hitler and his Nazi "super race" followers.

Standard Oil of California discovered black gold under the Saudi desert, which transformed the Kingdom of Saudi Arabia into one of the richest nations on Earth.

The Great Depression saw Franklin Delano Roosevelt re-elected by a landslide as the 32nd president of the United States, and Timberline Lodge on Mt. Hood, Oregon – where I have stayed and snow skied – began as part of a marvelous Great Depression Works Project Administration scheme.

In Great Britain, King Edward VIII abdicated the throne in order to marry Mrs. Wallis Simpson, American-born divorcée.

Finally, some oddball occurrences worth noting included the death of the last Tasmanian Tiger that had been held in Australian captivity, the creation of the Oscar Mayer Weinermobile that was created in the shape of a giant hot dog (weiner), and the name Spam was coined, a contraction of "SPiced" and "hAM."

The Oscar Mayer Weinermobile is still rolling today - "a hot dog on wheels" - was created in 1936 as a promotional stunt

SPAM, the food

Part One

The Early Years and Decisions

— Chapter 1 —
April 13, 1970: I Was Thirty-Four Years Old...

It began with a brief crackling noise on the downlink and then the terse crew report:

"Houston, we've had a problem!"

Warning lights illuminated all across the "eyebrow" of my EECOM console snapping me to attention. I examined more than 200 individual spacecraft systems' readings displayed on the two TV monitors to try to figure out what could have initiated so many lights. Little of the data made sense! In fact, so many of the parameters indicated failures of the spacecraft's systems that at first I believed it to be an instrumentation failure, such as I had witnessed with EECOM John Aaron when *Apollo 12* was struck by lightning soon after launch. I told myself: "It just has to be an instrumentation system problem, which should be able to be solved with the simple throw of a spacecraft switch as on *Apollo 12*," and with that thought I relaxed a little.

Nevertheless, I gripped the console TV monitor handles tightly, the ones we flight controllers referred to as "security handles," and stared intently at what the data were telling me: One entire critical cryogenic oxygen tank was gone, the other was leaking rapidly, and two of the three fuel cell electrical power generators had failed. It dawned on me that all the spacecraft equipment drawing power from these two fuel cells was now dead.

It was not an instrumentation problem, after all.

This realization was quickly followed by the thought that if the failures were unrecoverable, then the CSM would be lifeless in less than three hours. It was a monstrous failure and as EECOM it had landed squarely in my lap. Flight Director Gene Kranz, seated at his console behind me, and Astronaut CapCom Jack Lousma, whose console was separated by a two foot gap to my left, pressed me for recommendations, but I had no quick answers. My strategy was to find the leak and isolate it. It was an incredibly tense time; I was on the spot with absolutely no one to turn to for advice. The cold chill of panic began to rise in my body as a gorge, threatening to overwhelm me. It took a determined effort to not run from the MOCR in an unreasoning fit of flight.

It was not the first time that I had felt such an urge, however...

— Chapter 2 —
March 12, 1942: I Was Six Years Old…

Upon walking through the front door into our home in Kingston, New York, I was alarmed to hear my sister Phyllis, age nine, scream for me to run. What? Why? To my horror I saw that our mother, Ida, was brandishing a large butcher knife. She was chasing Phyllis, but upon spying me turned to chase me instead. Screaming, and in a blind panic, I followed Phyllis into a bedroom and we both dived under the bed. It was difficult to appreciate, but our mother's mental state had slipped over the edge, and she was now in the throes of a full-blown nervous breakdown. My instinct was to flee, but I couldn't.

> *Fear and panic came as an unwelcome visitor early in my life. I would not again experience those strong emotions until April 13, 1970.*

— Chapter 3 —
February 15, 1936: Born

The early 1900s were years of turmoil. They gave rise to the Bolshevik Revolution, anti-Semitism, and a war of unprecedented scale that overwhelmed Europe. Between them, they spurred a flood of East European immigrants to North and South America. In 1900, my paternal grandparents emigrated from Kiev, Ukraine, and in 1913 my maternal grandparents (Soloviov) brought their family from Novosibkov, Russia. Soloviov was Americanized to Soloff. My mother Ida Soloff, age four, was part of a large family of immigrants.

On August 16, 1930, 21 year-old Ida Soloff and Solomon (a.k.a. Sol) Liebergot, aged 20, eloped to Maryland. Her family was aware of my father's gambling reputation, and they did not approve of the marriage, so he had to talk her into running away. Nevertheless, upon their return, Bubba Pearl, Ida's mother, arranged a formal wedding and a beautiful wedding portrait.

I was born of Russian Jewish immigrant origins in a hospital in Camden, New Jersey on Saturday, February 15, 1936, at 9:40 p.m., weighing in at more than eight healthy pounds. I was named Seymour Abraham Liebergot, a first name that would cause me no end of teasing as a child, but I found I could run faster than the bullies who were chasing me, so I managed. To this day, I tell everyone that even with a name containing those twenty-three letters, I'm sure that my mother loved me – really!

In 1933, when Ida was pregnant with Phyllis, one of my father's gangster friends invited her for a ride in his new car. She was sorely tempted, but declined for some reason that she never explained. Whatever it was, Seymour Abraham Liebergot almost never was. Her intuition had been sound, because the gangster's car was blown up that same day and he was killed. A powerful bomb had apparently been wired to the ignition.

I had dodged my first bullet.

1930: Sol and Ida Liebergot formal wedding portrait

Sy Liebergot ★ 19

No. 54924

UNITED STATES OF AMERICA
NATURALIZATION SERVICE
ORIGINAL

PETITION FOR NATURALIZATION

To the Honorable the District Court of the United States for the Eastern District of Pennsylvania:

The petition of *Philip Soloff* 703 N. 4 St., Philadelphia, Pa., hereby filed, respectfully showeth:

First. My place of residence is ___ (Give number and street).
Second. My occupation is *Suit Presser*.
Third. I was born on the *25* day of *Dec*, anno Domini 1*871* at *Novatkar, Rusia*.
Fourth. I emigrated to the United States from *Liverpool, Eng* on or about the *10* day of *Feb*, anno Domini 19*13*, and arrived in the United States, at the port of *Phila, Pa*, on the *17* day of *Feb*, anno Domini 19*13*, on the vessel *Haverford*.
Fifth. I declared my intention to become a citizen of the United States on the *10* day of *May*, anno Domini 1*920*, at *Phila, Pa* in the U.S. Court of *Phila Co*.
Sixth. I am ___ married. My wife's name is *Jenny Russen Novatkar*; she was born on the *15* day of *Feb*, anno Domini 1*871* at *Russia* and now resides at ___.
I have ___ children, and the name, date and place of birth, and place of residence of each of said children is as follows:

Henry Mar 12, 1893 Russia Phila Pa — Rutie July 5, 1909 Russia Russia
Lewy May 2, 1894 Russia — Rutie May 27, 1913 do do
Sarah Aug 3, 1896 Phila Pa
Jacob Feb 14, 1899 do
Isaac April 17, 1903 Russia — Ida
Hyem Alle June 11, 1906 do do

Seventh. I am not a disbeliever in or opposed to organized government or a member of or affiliated with any organization or body of persons teaching disbelief in or opposed to organized government. I am not a polygamist nor a believer in the practice of polygamy. I am attached to the principles of the Constitution of the United States, and it is my intention to become a citizen of the United States and to renounce absolutely and forever all allegiance and fidelity to any foreign prince, potentate, state, or sovereignty, and particularly to *the Russian Government of Russia*, of whom at this time I am a subject, and it is my intention to reside permanently in the United States.

Eighth. I am able to speak the English language.

Ninth. I have resided continuously in the United States of America for the term of five years at least, immediately preceding the date of this petition, to wit, since the *17* day of *Feb*, anno Domini 19*13*, and in the State of Pennsylvania, continuously next preceding the date of this petition, since the *17* day of *Feb*, anno Domini 19*13*, being a residence within this State of at least one year next preceding the date of this petition.

Tenth. I have not heretofore made petition for citizenship to any court. (I made petition for citizenship to the ___ Court of ___ at ___ on the ___ day of ___, anno Domini 1___, and the said petition was denied by the said Court for the following reasons and causes, to wit, ___, and the cause of such denial has since been cured or removed.)

Attached hereto and made a part of this petition are my declaration of intention to become a citizen of the United States and the certificate from the Department of Labor, together with my affidavit and the affidavits of the two verifying witnesses thereto, required by law. Wherefore your petitioner prays that he may be admitted a citizen of the United States of America.

Philip Soloff
(Complete and true signature of petitioner)

Declaration of Intention No. *29813* and Certificate of Arrival No. *Not Given* from Department of Labor filed this *2* day of *Aug*, 19*22*.

AFFIDAVITS OF PETITIONER AND WITNESSES.

UNITED STATES OF AMERICA, } ss:
Eastern District of Pennsylvania,

The aforesaid petitioner being duly sworn, deposes and says that he is the petitioner in the above-entitled proceedings; that he has read the foregoing petition and knows the contents thereof; that the said petition is signed with his full, true name; that the same is true of his own knowledge except as to matters therein stated to be alleged upon information and belief, and that as to those matters he believes to be true.

Philip Soloff
(Complete and true signature of petitioner)

Harry Soloff, occupation *Tailor*, residing at *710 Rau St*, Philadelphia, Pa.,
and *Simon Schaffer*, occupation *Selectric*, residing at *215 N 6 St*, Philadelphia, Pa.,
each being severally, duly and respectively sworn, deposes and says that he is a citizen of the United States of America; that he has personally known *Philip Soloff*, the petitioner above mentioned, to have resided in the United States continuously immediately preceding the date of filing his petition since the *1* day of *Aug*, anno Domini 1*917*, and in the State in which the above-entitled petition is made continuously since the *1* day of *Aug*, anno Domini 1*917*, and that he has personal knowledge that the said petitioner is a person of good moral character, attached to the principles of the Constitution of the United States, and that the petitioner is in every way qualified, in his opinion, to be admitted a citizen of the United States.

Harry Soloff
(Signature of witness)
Simon Schaffer
(Signature of witness)
[SEAL]

Subscribed and sworn to before me by the above-named petitioner and witnesses in the office of the Clerk of said Court at Philadelphia, Pa., this *3* day of *Aug*, anno Domini 19*22*.

W. A. Uselyn, Clerk.

[OVER.]

1922 Soloff Petition for U.S.A. Naturalization

20 ★ APOLLO EECOM: Journey of a Lifetime

Sol and Ida Liebergot married ten months

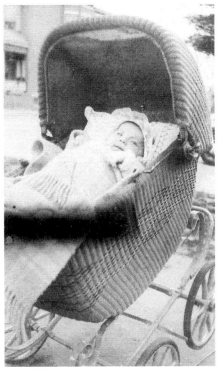

1936: Baby Sy Liebergot at age 3 months

1936: Sy at 11 months, Phyllis, and Mother Ida Liebergot

— Chapter 4 —
1939: The Journey Begins

My father was employed as a salesman by Pep Boys automotive store, in Camden, New Jersey. However, he had friends who were gangsters and unfortunately, gangsters who were not his friends – most notably the bookies and gamblers to whom he was forever in debt. Either way, he made a hazardous choice of friends and acquaintances.

Although there was insufficient money to buy food or pay bills, Ida suffered in silence, and life progressively worsened for us as Sol squandered what money there was on his drinking, gambling and philandering. Many times, when she suspected there was no food in our refrigerator, Bubba Pearl traveled down from Philadelphia, with bags of food.

Finally, Sol's gambling debts mounted to the point that he felt his life was in danger and he took desperate action: *he fled.* After stealing a sum of money from Pep Boys, he gathered my mother, sister, and me, and headed down the Eastern Seaboard on a bus.

We were on the lam (a gangster word for 'run') and it would not be the last time. The passage of time has blurred my memory of our various hideaways, but Raleigh, North Carolina was one, and the last was a tiny oceanside town with a boardwalk where we lingered several months. I recall there was a large ditch filled with muddy water where I had my first encounter with a crawfish. Ouch! I also recall being in a large drug store with my mother when I heard, "Call for Philip Morraise!" I spied a figure in a hotel bellboy uniform complete with pillbox hat walking around repeating his call. It was Johnny Roventini, the famous Philip Morris cigarettes bellboy.

My life had become one with no playmates and no certainties.

1931: Sol Liebergot (middle) in front of Pep Boys automotive store in Camden, New Jersey

Philip Morris Bellboy

— Chapter 5 —
1941: Philadelphia Memories

When it was safe to come out of hiding and return north, the destination was Philadelphia, Pennsylvania. We moved into a walkup apartment in a business district in West Philly. However, our stay there was not to be a long one.

Long road trips in the back seat of a car were so boring. One of the wonderful diversions available to us kids besides counting telephone poles were the Burma-Shave signs. This was the world's first brushless shaving cream and beginning in 1925 the company set out to win over customers with roadside signs that were the precursor to modern billboards on America's earliest roads. They were humorous jingles that were placed at intervals along the road, with each small sign showing one line of a four-part rhyme with the final sign concluding with the words "Burma-Shave." One such favorite was:

Don't stick your elbow
Out so far
It might go home
In another car.
Burma-Shave

Another, beard-related one stated:

Henry the 8th
Sure had trouble:
Short term wives
Long-term stubble!
Burma-Shave

I was now five years old, so I entered Kindergarten. I have wonderful memories of playing with building blocks, the cookies and milk snacks, painting with a watercolor kit, taking a nap on my little mat, and learning to ask for permission to go to the bathroom.

I looked down at my first lace-up shoes and my kneeling mother, as she patiently demon-

1941: Sy, 5 years old and sister Phyllis, age 8

strated the seemingly impossible task of teaching me to tie the laces in a special manner that had no name.

When I lost my first baby tooth, I placed it under my pillow in the confident belief that the Tooth Fairy would pay me a nocturnal visit, and I was duly rewarded with a dime. I also believed in the Easter Bunny and Santa Claus, but I wasn't sure who Santa Claus was.

In those days before television, my main entertainment was going to the movies. Ida gave me ten cents and I walked the short distance down the street to the local movie house, where I could barely reach the ticket window. The original *Mummy* movie was burned into my memory at the tender age of five, and provided me with nightmares for many nights. The movies had an additional clever feature in the continuing weekly serial, when the hero was left in a life or death situation from which escape was seemingly impossible. Of course, the next week he would escape, only to be trapped in another impossible situation. My child's mind accepted that the outcome would always be a harrowing last minute escape, but that same child's mind allowed for the possibility he wouldn't make it!

I came down with chickenpox and my mother kept me in a darkened bedroom, and she being a good Jewish mother, fed me copious quantities of chicken soup – what else?

I don't ever remember being hugged or kissed by my mother, or my father.

— Chapter 6 —
1942: Kingston, New York

In Philly, Sol apparently fell back into his old gambling habits, and so after a year there we were on the run again.

This time, we ended up in Kingston on the Hudson River, 100 miles north of New York City. It was a small town of 29,000 that was steeped in Revolutionary War history and, for a time, was the capital of New York State. Kingston was a town of "haves" and "have-nots," and we joined the second category. My father showed promise by starting Kingston's first taxicab and limousine service because the rich folks wanted private transportation to the surrounding Adirondack Mountain resorts and down to New York City. He bought a couple of 1940's Buick limos (the gangster-style cars that we all knew from the movies). However, it only took a short while for him to ruin what could have been a thriving business by allowing bookies to operate from his office and hold card games there. Once again he plunged into debt and left us with scarce food on our table.

The Perry family of Kingston started a cab company right after ours failed and became very wealthy.

— Chapter 7 —
Elementary School: First Grade

My formal schooling began in No. 8 Public School, in First Grade – and of course, like many of my classmates, I cried when my mother left me alone in that school the first day.

I sat at my new desk and the teacher placed a long, narrow card with strange markings on it upright in a groove before me. I asked the teacher what the strange markings were (I couldn't read, yet). She replied gently, "That's your name." It was a long card, printed in large block letters, which said, "Seymour Abraham Liebergot." "All of it?" I asked incredulously, as I moved a finger along the length of the card.

In the dark early morning hours, the milkman left real glass bottles of un-homogenized milk on the front porch. During the winter months, when I opened the front door to fetch the milk, the sight of the three to four inch columns of separated frozen cream that rose out of the frozen milk fascinated me. My child's vivid imagination allowed me to see frozen faces topped with hats of paper bottle caps.

I remember the beginning of the lyrics of a 30's song ditty entitled *The Iceman*, which began, *"The man that comes to our house, every single day..."* Every other day, the iceman delivered a block of ice that he inserted into the top compartment of our wooden icebox, which had not yet evolved into a refrigerator. We kids would gather at his truck and beg him for a free piece of chipped ice. I have heard those pieces of ice referred to as "caviar for kids."

— Chapter 8 —
World War II: Signs of the Times

I remember
- My father was classified 4-F war draft status (not eligible to serve).
- Food and gasoline rationing.
- I bent a knife cutting my first horse meat.
- Saving string in a ball, saving "tinfoil" from gum wrappers in a ball, saving grease and saving newspapers, all to be recycled for the war effort.
- I saved War Savings Stamps for War bonds.
- V-J Day newspaper headlines and the celebratory firecrackers.

— Chapter 9 —
1942: The Destruction of Ida Soloff

Sol continued to drive my mother down her emotional spiral. On the verge of losing the taxicab company, he drank heavily and rarely came home. Ida kept her own counsel, as her life became more and more desperate, never asking for help. Bills were not being paid, and having sufficient food on the table was a rarity. Sol compounded matters even more by having an affair with a poorly educated, 22-year-old Italian barmaid waitress named Josephine Oliveri. (We later learned that she was slated to travel to Italy to be wedded there in a pre-arranged marriage to a cousin.) One day he came home and announced to our silent suffering mother that he was leaving her for Josephine, who was pregnant with my half-brother Marvin. The shock of my father's double whammy brought my mother's weakened mental state crashing down like a flimsy house of cards. Even at the age of six and a half, I was aware of the change in my mother's behavior; she was constantly in an hallucinatory state. Childlike and paranoid at times, she would enter a closet to dress, "because the judge and the mayor were watching." The final blow came on the day that she grabbed a knife and attempted to kill Phyllis and me.

Was this her way of delivering us from a dreadful existence?

Ida Soloff Liebergot had departed completely from reality and would never return. In 1942, she was committed for treatment to the Psychiatric Center in Middletown, New York. She was first diagnosed with Schizophrenic Dementia Praecox, Paranoid type, which was later changed to Hebephrenic type from which she never recovered.

The downward spiral that our father had begun for us continued.

MIDDLETOWN
PSYCHIATRIC CENTER

COMMUNITY CAMPUS
122 DOROTHEA DIX DRIVE, MIDDLETOWN, NY 10940
845-342-5511 FAX 845-342-4975
TDD 845-341-1896

Mrs. Phyllis Pulley
1566 Lindbergh Avenue
Roslyn, PA. 19001-1525

September 17, 2002

Re: LIEBERGOT, Ida

Dear Mrs. Pulley:

I have your request for information on your mother, Ida Liebergot. We had to wait to retreive the chart in order to answer your letter.

The information I can give you is:

Mrs. Liebergot was born in Russia on August 7, 1910.
She was married on August 16, 1930 to Saul Liebergot.
She was first diagnosed : Dementia Praecox, paranooid type and later changed to Hebephrenic type.

She was of the Jewish faith until her death on October 8, 1967.

I hope this of assistance to you.

Very truly yours,

[signature: Joan Henderson for]

Ms. Sheryl DeLaGarza, MHIA
Director, Medical Record Department

A NY State Office of Mental Health Psychiatric Hospital
Affiliated with Columbia University for advanced psychiatric training in Geriatric & Public Psychiatry
An Equal Opportunity/Affirmation Action Employer

— Chapter 10 —
Foster Homes

In the same year that my mother departed reality, Sol came to terms with his inability to both support his expensive life style and provide for two children, so he removed Phyllis and me from school and placed us in foster care.

Ironically and chillingly, this scene would be replayed eleven years later, with my three half-siblings.

My sister and I spent a number of months in a succession of foster homes, but at least we were together. Our first placement in Kingston was with a stern-faced middle-aged couple who provided a place to live and cared for us, but nothing more. The last, however, was a child's dream come true! We were placed with Grandmother Black, her son and his wife, who owned and lived on a farm in the rural community of Port Ewen, New York, a few miles southeast of Kingston. The farm featured a large house, barn, milk cows, horses, a truck garden, and even an apple orchard. It was idyllic. It was all absolutely perfect for a six year old boy with a vivid imagination. However Phyllis, older than I, and more aware of our plight, cried all the time for our mother.

Grandmother Black baked fresh bread every day, and the aroma was intoxicating. The house had a root cellar in which fresh and canned vegetables, homemade jams and preserves were stored. It was fascinating to watch her prepare the preserves from berries picked earlier that day, finally pouring on the melted wax to seal the delicious product and screw down the cap on the classic Ball jar. Ah, fresh homemade bread and preserves – could life be better?

One day I approached Grandmother Black who was methodically moving what appeared to be a broomstick up and down in a vertical tub. She invited me to spell her at the task, which I obediently did. When I asked what we were doing, she informed me that the device was a butter churner and we were making butter!

Phyllis and I were enrolled in the Port Ewen School; the community was so small that a one-room building housed grades one through eight. I was in the front of the room continuing in First Grade, while Phyllis was further back in Grade Three. I noted with juvenile amusement that the physical separation between First and Third Grades was only about eight feet.

I treasure the unique memory of that one-room schoolhouse.

I was a bed-wetter and a thumb-sucker; I suppose nowadays I would be labeled as having serious emotional problems. Today the cure would most likely be gentle psychotherapy, to get to the root cause of my problems, which would be explained using psychobabble. Grandmother Black took a more direct approach to the bed-wetting problem: she saved the wet sheet for me each day and rubbed my face in it when I returned from school. I knew trouble was coming because she would hang the sheet out of the bedroom window so I would see it. In our more enlightened times, this would be classified as child abuse, but it worked for me, curing me of bed-wetting, and seems not to have left any emotional scars.

Grandmother Black's son was a disciplinarian. He and some men spent hours one day, emptying, cleaning, and refilling a large circular wooden water cattle trough. As a feckless seven-year old, I rewarded their effort by emptying a bucket of red barn paint into the freshly cleaned trough. To this day I am clueless why I did so, nor do I recall the act, which my sister swears took place. However, what I do vividly recall was the thorough beating that I received. Ouch!

Early one morning I watched Mr. Black milk a cow, and I was transfixed by his rhythmic pumping of her teats and the sound of the alternating streams of milk impacting the bottom of the pail, which faded as the pail filled. He paused, telling me to come close and open my mouth. He squirted the raw milk from a teat directly into my mouth. Delicious! He then sat me on his stool and invited me to try my hand at milking Bossy. That first (and only) attempt at milking a cow produced no milk, and I was rewarded with a kick from the cow's hind leg, just forceful enough to knock me sprawling.

Phyllis and I had a wonderful surprise when an ambulance arrived one day, bringing our mother for a visit from the Middletown Psychiatric Center; I suppose this was an experiment to determine whether there would be an improvement in her behavior. Phyllis said she arrived in a straight jacket and kept repeating "My kindra, my kindra," (My children, my children) when she saw us. She was still childlike in her demeanor, and in joining us in the hay in the barn, she played as if we were playmates. Despite what she said, I don't believe that she really knew us. We did not see her again for five years, this time in the Middletown hospital, at which time she displayed absolutely no recognition of us.

That was the last time either of us saw her alive.

Despite my thumb-sucking and bed-wetting, and my sister's sad emotional state, Mr. and Mrs. Black wanted to adopt us and they initiated steps towards this end. However, upon learning of this, our father abruptly reclaimed us.

— Chapter 11 —
1943: Together, Again

Sol collected Phyllis and me, along with Josephine, who was pregnant with our soon to be half-brother Marvin, and moved us to a little out-of-the way place in the countryside northeast of Kingston, named Saxeville, Old Hurley. I've always wondered how he decided on the places to move us. In retrospect, they always seemed to be good places in which to hide out.

With Marvin, a dog and a cat we became a crowded family unit. The animals weren't too friendly – the white Spitz dog always bit me and the large tomcat bit and scratched me.

Our new home was located down a small country road that ended at a creek. It was perched on the tree-lined banks of the Esopus Creek, named after the native Indians who once lived and hunted there. Arrowheads and spear points were a common find in the surrounding fields. The small two-story house had a coal-fired furnace for heat, a wood stove for cooking, and the small, clear creek provided drinking water. It was fascinating to watch Josephine lift a cooking plate and put small pieces of wood into the stove while cooking a meal.

I've always wondered: How do you 'simmer' on a wood stove?

Wintertime was a joy for me, though most of the time I had to play alone. With snow on the ground and the creek frozen, I improvised a sled and slid from uphill, gathering speed until I shot down the creek bank and across the ice to its other side – Exhilarating! However, winter and the coal furnace gave real meaning to an old saying of "hauling the ashes."

A neighbor man who lived uphill just next door to us was forever tinkering with a long conveyor-looking contraption of belts and what seemed to be large roller brushes. I finally mustered the nerve to visit in order to watch him adjust 'this,' and tinker with 'that' on the machine. He completely ignored me. When he picked up a bushel of apples and dumped them in one end of the machine and they disappeared in a sequence of whirring, clacking, and spinning brushes, curiosity got the better of me and I asked what he was working on. He stopped his activities and told me that he was inventing an apple-polishing machine. And sure enough, as I watched with fascination, shiny apples exited the other end of the long machine. I've often wondered if that was the genesis of apple-polishing machines.

The nearest school, Hurley Public School, was a mile away, and was reached by bus or on foot (if you missed the bus). Phyllis and I began school once more, I again in First Grade for the third time in one year. As the only Jewish kids in that school, we experienced overt aggressive anti-Semitism for the first time in our short lives.

I didn't understand – what's a Jew?

— Chapter 12 —
1944: Kingston, Part II

After a year of austere creek-side living, we moved back to Kingston. Sol took a job as a knitting machine operator in the Barclay Knitting Mills, one of the largest mills in the state. As events unfolded, we remained here for the next five years before he had to take us on the lam … *again.*

— Chapter 13 —
Second Grade: A Lasting Impact

When I began second grade in No. 6 Public School, I encountered "Miss Mean Teacher" (for the life of me, I can't recall her name). She was middle-aged, simply dressed, stark and plain in appearance, with her hair stereotypically tied in a bun. I was finally able to sound out words and was actually reading a little. I could finally read my name – but that was not much of a gift. However, I do recall that I was proud that I could successfully spell and sound out 'restaurant.'

The boys would kiss up to Miss Mean Teacher by volunteering to clean the slate blackboard after school. I was not immune. Another boy and I volunteered for the duty one winter afternoon, washing the blackboard and dusting the erasers.

The next morning when I arrived in the classroom, Miss Mean Teacher called me to her desk at the front of the room and accused me of stealing two 10-cent War Savings stamps from her desk. I indignantly denied this, and suggested that it must have been the other boy, but she would have none of it. I guess I must have looked guilty. After the whole class assembled, she stood me in front of my classmates and informed them that I had stolen two stamps from her desk, and I was a thief and a bad person. She then marched me to the principal, who interrogated me even more vigorously, pressing me for an admission, but I adamantly maintained my innocence.

The ultimate injustice was that the matter was reported to Josephine, who unsympathetically believed the school and so I had to 'repay' the 20 cents, which I did by shoveling snow from the sidewalks of a couple of neighbors.

When Valentine's Day rolled around the next week, we were obliged to exchange cards. We all waited with eager expectation to see who would garner the most cards. As they were passed out, it began to look as if I had received no cards, I quietly began to cry, the salty tears sliding down my cheeks. I sat with my head bowed down, until I sensed someone approaching. It was a girl, holding a single card in her hand. I raised my tear-stained face as she laid it on my desktop in front of me. Neither of us said a word, and she turned to return to her desk. The act was not repeated by any other student. I wish I could tell her what that tender gesture of kindness meant to me then and, in fact, still means to me today. Of course, the rest of the class had been watching in rapt attention and you could have heard a pin drop. However, Miss Mean Teacher totally ignored the tender event.

I suppose there is a lasting emotional scar from this episode in that, in later life, I always overreacted whenever I thought I was being accused of something I had not done. It became a "hot button" for me.

As far as the memory of Miss Mean Teacher is concerned, I am sure that I shall never remember her real name.

— Chapter 14 —
Stability ... At Last?

My father, unable legally to divorce Ida, could not marry Josephine, but that didn't stop him from siring more children. Josephine was now pregnant with David and we were soon to expand to a family of six as we moved into the first floor of the small two-story wood-sided building at 100 Pearl Street, in a pretty, middle- to upper-class part of town. The Lessick family lived on the second floor. Irv Lessick, a year older than me, soon became my best friend and he adopted me into his circle of chums, one of whom was a cousin whose father was an owner in Barclay Knitting Mills. The cousin's family was rich, and they lived in a 23-room mansion up at the end of Pearl Street and owned a brand new Oldsmobile. It was my first exposure to real wealth.

The first seeds of a desire for a better life goal were planted.

I continued my elementary school education in Third Grade at yet another school, No. 7 Public School, where I was able to complete Seventh Grade, and go on to Myron J. Michael Junior High school for Grades Eight and Nine.

Third Grade became significant for my personal hygiene. I was bereft of any parental guidance in personal hygiene, particularly in teeth care and rarely brushed my teeth. My teeth suffered sorely, which prompted my concerned teacher to lecture me on proper teeth care and send me to the school nurse. She arranged for free dental care through the school after determining that my parents could not afford to pay a dentist. He was a marvelous dentist and after somehow convincing me that Novocaine was not necessary, repaired nineteen teeth without that local anesthetic

1948: Kingston, NY L-R David, Phyllis and Marvin Liebergot

(I have a suspicion that he somehow hypnotized me). Unlike today's high-speed drills, he was limited to the use of the common low-speed drill that was driven by a belt with a dark joining splice. That splice caused the drill to be characterized as "chase the rabbit" and that splice was all I could see while he worked on my teeth.

I still had no idea how poor we were.

When I turned ten, I landed a job as a newspaper delivery boy with the *Kingston Daily Freeman* newspaper. I had 43 customers, and was paid $2.50 per week from which my father took a dollar for "room and board," he said. I felt that I needed a bike for my paper route, so I saved up and was able to buy a used Columbia bicycle from one of my customers for $15. My route was very hilly: I don't know if peddling that heavy bike was easier than walking, but it was my first possession and I was proud that I had purchased it by myself. The newspaper was a thin one, so I could fold it into a flat square shape that was perfect for throwing; if my aim was on, it sailed onto my customers' porches, but if it was off, it could easily end up on their roofs.

100 Pearl Street, Kingston, New York, 1950

— Chapter 15 —
Babies, Beatings, Pasta

Most of the time we had one meal a day, and sometimes the dog lost out because I ate his biscuits which, when you're really hungry, are pretty tasty. It seemed that all Josephine ever made were meals consisting of various combinations of pasta and peas: spaghetti and peas, elbows and peas, rigatoni and peas, shells and peas. Occasionally, she made lentil soup. Sometimes, as a real treat, she would make a large roasting pan of baked macaroni and cheese (Hail Velveeta!). I'd also go out into neighbors' yards to pick dandelion leaves for a vegetable – then a weed, but now considered a gourmet food item. Hungry, and alone at home one winter afternoon, I was inspired to scoop up a glassful of snow and poured pancake syrup over it. Snow cone *au natural*? I tried it on another occasion with vanilla extract. How could something that smelled so wonderful have tasted so horrible?!

My father continued to work as a knitter, and Josephine took a job as a waitress to help sustain the larger family, which placed great stress on her because the long, hard hours at work were in addition to keeping house. Her endurance apparently had reached the breaking point when Sol discovered her in bed completely unresponsive – she was dying from an overdose of sleeping pills. As Ida before her, Josephine sought ultimate relief from her desperate existence – one by leaving reality, the other by killing herself. Sol yelled to me to come in the bedroom and we lifted her limp body from the bed, walking her around the room to keep her awake until the ambulance that he'd called had arrived.

> *How ironic that Sol had driven one woman out of her mind and another to attempted suicide.*

I watched Josephine, on her knees, bending over the bathtub, strenuously washing clothes on a scrub board. Finally, we were able to purchase a marvelous device called a washing machine, constructed of a cylindrical tub and a set of wringer rollers mounted on top. One day, Marvin decided to investigate the workings of the new machine. I heard him screaming, and ran to investigate, whereupon I discovered him with one of his arms through the wringer rollers, up to his armpit. Fortunately, he suffered no lasting damage, only a lasting lesson.

> *Our life was falling into the same pattern that Sol had inflicted on Ida. There was no time spent on those things that form close family bonds: shared activities, attention, and affection.*

I suffered a lot of beatings by my father, who seemed to relish the opportunity to use his belt and belt-buckle on me. These were real beatings, during which he really worked up a sweat as he seemed to vent his own frustrations. Sometimes, for additional punishment, all my clothes were taken from me and I would be locked in my bedroom, naked. That was Josephine's concept of "grounding." Undeterred by my nakedness, I would slip out the bedroom window, run through neighbors' back yards until I reached a friend's house where I was able to borrow some of his clothes and off we went to play. I must have been quite the sight while hopping fences totally naked. I would return through the window into my bedroom before Sol came home. "Wait till your father gets home!" had real meaning when Josephine threatened it. She didn't like me; I was just a smart-mouthed nuisance. Frequently she would curse me in Italian with phrases that meant "You're crazy" and "Hard head!"

In today's world of excessive government intrusion into our lives, a government Child Protection Agency probably would have taken me away and placed me in a foster home. Or I may even have been able to sue my father for a divorce from the family. Hell, like many of my generation, I suffered the humiliation of being bent over the school teacher's desk to be given a few whacks with an eighteen-inch wooden ruler, although in that case, the punishment had been deserved: I had dipped one of Shirley Terwilliger's pigtails in my desk inkwell.

1950: Kingston, New York, Sy Liebergot and sister Phyllis

Meanwhile, Phyllis betrayed her growing neurosis by twisting and pulling out her thick dark brown hair, strand by stand, until ultimately she became essentially bald and had to start wearing a wig.

As for me, I was fast developing an emotional callus that would hinder me from forming meaningful relationships for years to come.

At some point I realized that I no longer cried. Progress?

— Chapter 16 —
1945-1950: Signs of the Times

- I used to sit on the floor with my ear against the big Philco console radio and listen to my beloved Brooklyn Dodgers get beat by the hated Yankees in the '47 and '49 Baseball World Series. We beat them in the '55 World Series, but the revenge was short-lived.
- I recall that our telephone number was only 347 and it was a party line.
- The traffic lights signaled only red and green.
- There were no traffic "Yield" signs.
- Two inches of snow fell in July.
- I melted lead and made toy soldiers.
- I made firecrackers with two bolts, two nuts and kitchen match heads sandwiched in between.
- I saw an Indian motorcycle parked on the street.
- Cherry trees and rhubarb grew in a neighbor's yard.
- We actually were able to go to a park and play baseball without being organized into leagues.

— Chapter 17 —
Growing Pains

The Magic of Reading:

Most times I played by myself, alone with my comic books and homemade toys. As long as I didn't get into trouble, I was left to my own resources. I soon discovered science fiction books and magazines; and with my young boy's rich imagination, I was transported far away from the confines of my small bedroom. Having discovered the magic of reading, I read voraciously, expanding both my vocabulary and my horizons.

I first discovered Edgar Rice Burroughs' novel, *John Carter of Mars*, which was full of swords and ray guns. After reading H.G. Wells and Robert Heinlein, I was hooked, no doubt about it. I spent my teens reading pulp magazines with names such as *Amazing Stories*, *Astounding Science Fiction*, and *Galaxy Stories*, which introduced me to the likes of Isaac Asimov, Ray Bradbury and Clifford Simak. I feasted on the salmagundi served up by the collaborations of C.M. Kornbluth and Frederik Pohl and on the humor and satire penned in short story form by Frederic Brown and Robert Sheckley.

And of course, I will never forget the goose bumps I experienced later in life when the alien Black Monolith was shown for the first time in the lunar excavation scene in the Stanley Kubrick's movie adaptation of Arthur C. Clarke's, *2001: A Space Odyssey*. I am still a science fiction junky.

My Criminal Career:

At the age of 10, I discovered that it was easy to go to the Kress 5 & 10 store and casually shoplift items that *I imagined* I needed. Pretty soon I was doing this in all the stores. When I finally got caught lifting cans of food from the Grand Union food market, the store manager reported me to Josephine by telephone and then let me go home. I wasn't so much concerned about getting caught as I was frightened at the prospect of my stepmother issuing that awful threat, "Wait till your father gets home!" When he did, I received another brutal beating, although on this occasion, I knew that I had it coming.

The next and last episode of my criminal career came when a couple of my chums found a way to get into the basement of the local 5 & 10 department store and invited me to join them one Sunday afternoon. It was exciting to wander around in that basement wonderland among all the shelves full of goodies. However, we got found out the next day when one boy confessed to his mother, and I stood quaking before the store manager listening to him loudly threaten to turn me over to the police and "have them lock me up in the jail deep under the police station." He so thoroughly frightened me (I'm surprised that I didn't pee in my pants) that I vowed then and there, never, ever, to repeat my poor behavior. Amazingly, I reformed myself without receiving a beating. Thank heavens for that measure of tough love from the store manager.

My days as a career criminal were over.

Loss of Vision:

When I was 12, I awoke one morning to find that a dark veil and black floaters had appeared in my left eye, severely impairing the vision in that eye. I was frightened. I became even more distraught when the local eye doctor said that he could do nothing. I learned to live with the disability, and finally ignored it. In later years, it was not a hindrance to my sports endeavors: I excelled in gymnastics, track / pole vaulting, springboard diving, and snow skiing.

Just another challenge.

Born extremely nearsighted, I have observed that vision correction is a love / hate vanity process occurring in phases:

Age 10: I *hated* having to wear glasses for the first time.
Age 37: I *loved* my first pair of contact lenses.
Age 45: I *hated* wearing my first bifocal contact lenses, but I developed "eyelids of steel."
Age 65: I *loved* to be able to wear my first pair of "progressive lenses" glasses.

Should I be Jewish?:

After attending the Bar Mitzvahs of several chums, mainly for the free food, I figured that since I was Jewish I had better be Bar Mitzvah'd as well, so at the age that most Jewish boys were being confirmed in the religion, I enrolled myself in Hebrew School. It was a bit embarrassing for the teacher, as the other beginning boys were only seven. I soon had to drop out, however, because I had insufficient money of my own to pay for the schooling.

So I wasn't going to be Jewish ... I couldn't afford it.

— Chapter 18 —
Kingston High School: Music Appreciation

Although Kingston was a small city with only one junior high and one high school, it boasted an á cappella choir of some statewide fame.

My Aunt Helen Liebergot told Phyllis that our mother could play piano by ear, making her quite popular as a young girl, and at one time we actually had a piano. I remember none of that, though, so my first conscious exposure to live music came when I moved up to 10th grade in Kingston High School and heard the á cappella choir rehearsing. The power of more than 100 voices singing was so overwhelming that I found myself yearning to be part of it. Shaking with a mixture of excitement and nervousness, I auditioned for the choir's director and, to my surprise, I was accepted. Fortunately, my voice had made a smooth transition from soprano to tenor as I entered post-pubescence. I began singing lessons that were provided by the school to a group of qualifying students; how I looked forward to performing my lesson in front of them. To my enormous disappointment, that came to an end after only a month because we were once again uprooted.

It wasn't until 40 years later, that I would feel compelled to take up singing again.

— Chapter 19 —
December, 1950: Back to Philadelphia – A New Culture

My father had disappeared. One day, I saw Josephine open the front door to a uniformed policeman who said that he was "looking for Sol Liebergot." She replied that she didn't know where he was. Actually, he had fled to Philadelphia to avoid paying the debts, both legal and illegal, that he had accrued. Soon, Josephine loaded all of us on a bus and we headed down to Philly.

We were on the lam, again.

Having previously lived in Philadelphia, Sol had readily found a job as a cab driver, and a place for us to live, this time in North Philly, at 17th and Venango Street, a few blocks from the large intersection of Broad, Erie, and Germantown Avenues. We moved into an apartment on the second and third floors above a corner neighborhood food market, so I was able to secure a part-time job there after school.

My half-sister Denise was born the next year, 1951; now there were seven of us. We were definitely not the Seven Little Foys, but we seemed to frequently "go on the road."

The culture shock of the Big City overwhelmed me. While most businesses in the neighborhood were (still) owned by whites, most of the Jewish residents had fled "to avoid living next to the schvartzes" (Blacks) and the residences were now ceded to those new owners. My new neighborhood was a black ghetto about which I understood nothing. I was totally naïve about racial prejudice, as my contact with Negroes (the label Negro was the word of that time) in the little town of Kingston had been little to none. My 'first contact' was finding myself surrounded by a gang of ten small Negro boys, carrying broomsticks, and I was cowed. However,

1950: Sol Liebergot

after a few minutes of intimidation involving bullying verboseness and stick waving in my face, they departed, leaving me shaken. Later, after I had made new friends in the neighborhood, I was able to laugh at the incident.

So Seymour, who was the new kid on the block, fourteen years old, skinny and white, complete with glasses, set out to make friends in the new foreign land. Fortunately, Junior Richardson, who was black, and five years older than I, took me under his tutorial wing and protected me as I learned the ropes. I needed all the help I could get since I was so skinny and nerdy-looking. He became my best friend. Junior took me into his family; there was no father, but a wonderful mother who earned extra money caring for foster children, to whom I could certainly relate. Through Junior, I met and was accepted by others who enriched my life with their culture and allowed me to mature in an environment that would stand me in good stead the rest of my life.

Was Junior Richardson assigned to be my 'Angel'?

I learned that I had to stand up and fight (with fists) whenever a dispute arose, which in my case was frequently; there was always someone to challenge me over something or other, and I had to be ready to "put 'em up!" I also had to learn "jive talk," the vernacular of the neighborhood. Such was the ghetto protocol. I learned never to mess with "Little John"; the Little Johns of the neighborhoods were the strong, quiet ones who would kill you rather than look at you.

You didn't "spit into the wind" and you didn't mess with Little John.

It's fortunate that my Philadelphia ghetto experience preceded the common use of knives and guns, or else I believe that I would not have survived.

One of the few times that Sol showed some concern for my welfare was in the selection of the high school for me to attend. The nearest was Simon Gratz High School, only a few blocks away, and I was destined to attend there until my father discovered that the 100% black school had a reputation for being very rough and hence no place for me. Through his efforts I enrolled at Northeast High School, which was outside our district and a twenty-minute trolley car ride away. Junior Richardson later confirmed to me that if I had attended Simon Gratz, it would have been "detrimental to my health," because even the girls were said to carry knives hidden in their hair.

I had dodged another bullet.

Photography as a hobby bloomed for me at age 15, as I went from door to door, asking if I could take pictures of their children. Hustling for home portraiture added income to that from my part-time food store job. My new wealth allowed me to buy nice clothing and, at age 17, my prize, a royal-blue 1940 Ford Deluxe Sedan for which I paid the handsome sum of $100. I now owned a car and my father did not. He actually asked to borrow it on occasion. Of course I consented!

Through my photography, I met my next best friend Irv Lipkin, who was white, Jewish, and 10 years older than I, as were all his friends. He still lived at home at the age of 26. Home was behind and above the family business, a tailor shop where Irv, his mother, his father the tailor, and his older brother who worked in a dye factory, all lived. Irv wanted to lead an independent life, but they discouraged him from leaving home. He left a couple of times, but each time his brother hired a private detective to find him and convinced him to return. He lost the will to resist, and so he stayed.

A wasted life.

On the positive side, through Irv and his friends' examples I began to learn how to act as an adult. We both loved hi-fi and photography. He had been an electronics technician in the Army and loved to build little audio amplifiers and he had a complete darkroom in the cellar where we spent many an hour. I had such a passion for photography that I dreamed of becoming a photojournalist for *National Geographic* or *Life* magazine. I habitually wandered around the city and its parks snapping pictures of people, objects, and situations, which sharpened my eye and skills. And I even received an occasional assignment from neighborhood newspapers. It was a thrill to see a photo credit with my name!

— Chapter 20 —
December, 1950: Northeast High School

It was with some anxiety that first morning that I stepped off the No. 20 trolley car at 8th and Lehigh Avenues and got my first look at Northeast High School, an imposing ugly old stone structure made of granite blocks. A temporary Northeast came into existence in 1885 as an all-boys' public high school, and graduated its first class in 1892. The permanent granite structure at 8th and Lehigh was completed in 1905. In fact, there were two graduating classes each year. My class graduated in June 1953, as the 102nd. There was also an all-girls public high school, known as Girls' High, which continues in operation today. We didn't know then that today it would instead be called a "same sex" school, a concept that is regaining some popularity.

The school song was entitled *Hail, Northeast*, and the lyrics were sung to the melody of *Aura Lea*, an obscure love song written in 1861 by G.F. Poulton soon after arriving in the United States. In 1956, Elvis Presley's version, called *Love me Tender*, sold one million copies even before being released. *We were almost hip.*

Our flight to Philadelphia meant that I enrolled in the middle of the first semester of 10th grade. Fortunately, I was allowed to pick up as best I could, but I essentially lost the entire semester. The high school had teachers who acted as class advisors to assist students in establishing their schedule of classes for the coming semester. My adviser, Mr. Sullivan, assumed that I would take geometry, which I had missed in the prior short semester. However, he was taken aback when I told him, in no uncertain terms, that I wanted *no* math for the rest of high school, since I had done so poorly in geometry in Kingston and had lost confidence in my ability to comprehend math. He argued mightily with me and we finally struck a deal in which I would take geometry in the department head's class. Well, to my surprise, I earned an A. My discovery of the joys of math would drive my decision to go into engineering later in life.

I had dodged another bullet. (Another 'Angel'?)

With 2,500 students, Northeast High was only at half capacity. Approximately one-third of the boys were black. Although racial tension was a factor, and occasionally there were scuffles in the hallways, these incidents never lasted long. I took pains to steer clear of any of this because I had had my share of everyday confrontations in the totally black neighborhood in which I lived.

There were lots of situations to laugh at, as well. There was an elderly English teacher, and I mean OLD, who had such poor vision that he reminded me of the cartoon character, Mr. Magoo. Students were able to come in late for class by crawling on the floor and quietly slipping into a chair. However, this was thwarted one day when one of the students arrived late and fell to a crawl with a length of soft pretzels liberally slathered with yellow mustard (a Philadelphia staple). Well, the old guy may not have had very good vision, but he surely did possess a keen sense of smell and as the aroma of all that mustard wafted through the classroom, he demanded, "All right, who just came in late?!"

I joined the gymnastic team in 10th grade, and with the encouragement and training of the coach, Gustav Baack, I learned daring skills and developed some self-esteem in which I was sorely lacking. The lengthy meets were always against two other schools and were nerve wracking. A special treat came when the five best team members were selected to travel to the Annapolis Naval Academy in Maryland to compete against their Plebes (Annapolis freshmen). During my two years on the team, I was selected to compete, but the older Plebes wiped the mats with us younger high schoolers.

The initial trip to Annapolis produced my first incident of racial discrimination. When we got off the train from Philadelphia, we all headed to a nearby diner for soft drinks. Our small group of eager young athletes included Ben "Foxy" Wright, who was black, but to me he was just another teammate. However, the white guy behind the counter refused to serve Ben, telling him he had to remain outside. We were incensed, but Ben graciously insisted that we not make a scene, so we bought our soft drinks, including one for Ben, and left the diner to rejoin him.

In 11th Grade, in the absence of any parental pressure or encouragement, I did only what was necessary to get by, and my academic performance suffered. Nevertheless, I made a decision to "get with the program" because, even though I had no foreseeable plans to attend college, I wanted to protect that option. As a result, I went "all out for school" in my senior year. I reactivated the photography club, I became photo-editor of both the school newspaper and the yearbook, I was elected to the Northeast Honor Society, I was awarded the Northeast Senate Award (highest non-athletic award), and I was lettered in two sports: Gymnastics and Track (pole vaulting). At the end of my athletic career, I was 5 feet 8 inches short and 120 pounds of gristle. Academically, I finished in the top 10% of my senior class. Privately, I felt very proud, but I didn't share my feelings with anyone.

With my many extra-curricular activities, my high school days were hectic, and they became even more demanding when Sol decided to try again at owning a small business.

Sy Liebergot high school pole vaulter

This time he started up a diner in a small building next door to our apartment building, which was very convenient, as it allowed Josephine, my sister Phyllis, and me to work there. As Sol knew little about running a small business, he kept his job as a cab driver during the day, and helped out in his diner in the evening. By now, his cigarette habit had grown to five packs a day, so he was never without a cigarette in his mouth, which was hardly hygienic in food preparation. I worked in the family business every evening after school as a short-order cook and did my homework during slack periods, which grew longer as the initial rush of clientele fell off. I learned to make great hoagies and Philly steak sandwiches; the Philly Cheese version hadn't been invented yet. Sadly for Sol, but to my secret relief, the business venture folded after about a year.

Before graduation in June 1953, I petitioned Vice-Principal Howarth for help in getting a job at the *Philadelphia Inquirer* where I was eager to work as a newspaper photographer. He looked at me quizzically, asked me who I was, and expressed wonderment that he had never heard of me. I showed him my class yearbook, in which my picture appeared no fewer than eleven times. Satisfied, he made a call to an editor at the *Inquirer* and secured me a job there, albeit not the one that I had wanted.

When graduation day came I didn't tell anyone. No one in my family was particularly interested in any of my school achievements, so I rarely spoke of them. The day before, I rode the Broad Street Subway down to South Philly where, on South Street, there were 'schlockmeister' shops and pushcarts and I bought a secondhand suit so that I would look presentable at my graduation. Many years later I told Phyllis about my graduation and she was mortified that she had missed it.

In 1957, the all-male Northeast High School vacated its 8th and Lehigh location and a co-educational school was built at a new location. Soon after, a Medical, Engineering and Aerospace Magnet School was established as part of the main campus.

In 1996, I returned to the "new" Northeast to be inducted into the school's honored alumni "Wall of Fame" after a forty-three year absence. Afterwards, I had the pleasure of speaking to the 800 Magnet School students about my *Apollo 13* experiences.

Sy Liebergot 1953: Northeast High School graduation photo

— Chapter 21 —
Classmates

As I leaf through the yet-to-yellow pages of my fifty-year-old copy of *The Archive*, our graduating class yearbook at Northeast High, the senior pictures of three classmates clearly stand out in my memory.

Leon Slawecki was president of the Rocket Society. As school photographer, I did a photo cover for their newsletter. It is significant to note that Leon and his friends were "Rocket Boys" in 1952, a full five years before Homer Hickam Jr. and Sputnik. I was certain that Leon must have gone on to become a physicist as he stated in the yearbook. In May, 2002, using an internet search, I found him living in Washington, Virginia, doing "gentleman farmer" chores on 26 acres of land where, among other things, he grew grapes from which he made a decent chardonnay. He did not become a physicist because he realized that the math was beyond his ability. Instead, he went on to earn a Ph.D. in International Relations at Yale, and continued on to a successful career as a diplomat and international relations scholar, along the way becoming proficient in four languages, including Chinese.

Leon Slawecki 1953: Northeast High School graduation photo

Guy Rodgers was a basketball team star whose stated goal was to become a physical education teacher, but he far exceeded this by achieving NBA stardom as a professional basketball player with the Philadelphia Warriors. Sadly, Guy passed away on February 19, 2002, aged 65. As the school newspaper photo editor, I had the opportunity to see Guy's basketball skills up close and often. He was liked and admired by many classmates, including me. He and I knew each other well, since I would endeavor to provide him with good action photos from the games. I recall in 1961, eight years after graduation, waiting for him to emerge from the dressing room after a losing game between his visiting Philadelphia Warriors and the Los Angeles Lakers and wondering if he would remember me.

Guy Rodgers: 1953 high school graduation picture

He didn't disappoint me; as he came through the doorway, he took one look at me and said SEY-MOUR LIEBERGOT! What a surprise! He was that kind of "guy." Just then, Wilt Chamberlain exited the Laker dressing room and I got to shake his enormous hand. He was no longer "The Stilt." He had bulked up to a real brute since his Overbrook High School days playing against Northeast High.

Burt Zeldin was another best friend. He was a wild kid who habitually scored bad grades in high school. His stern locksmith father always kept a tight leash on him, which was probably why he was so wild when out of sight of his father. Burt had a special fascination for pyrotechnics … of his own making. It's a wonder that we didn't kill ourselves with his concoctions. One of our favorite pranks was to mix one of his special powders and place it as a small packet on a trolley car track. When the heavy streetcar rolled over the pyrotechnic powder it detonated, rocking the car just enough to cause the driver to stop and get out and look around. The puzzled look on his face was priceless. Years later, during my NASA career, I discovered Burt working as a scientist at the Jet Propulsion Laboratory (JPL). He had earned a Ph.D. in Heat Transfer from the Penn State. He is retired now, and we maintain contact with each other.

1961 basketball program from the Philadelphia Warriors vs Los Angeles Lakers game

Burt Zeldin 1953: Northeast High School graduation photo

— Chapter 22 —
High School Years: Signs of the Times

- Philadelphia scrapple, soft pretzels, hoagies and steak sandwiches.
- Riding trolley cars.
- The New Year's Day Mummers Parade.
- Wildroot Cream Oil hair cream.
- Ducktail hair style with a pompadour (very cool).
- White shirt, collar turned up, high riser pants with pegged pants (very cool).
- The Korean War breaks out.
- Ninety students were expelled from West Point for cheating.
- The U.S. Senate Kefauver Committee investigates organized crime for the first time.
- Movies: *Death of a Salesman* and *A Streetcar Named Desire.*
- Pop singers: Joni James, Patti Paige, Theresa Brewer, Frankie Laine.
- Dwight D. Eisenhower elected the 34th U.S. President.
- Joseph Stalin died.

— Chapter 23 —
1953: The Philadelphia Inquirer

After high school graduation I reported to the Philadelphia Inquirer newspaper for the job interview Vice-Principal Howarth had arranged. I found my way to the Photography Department and sat across a desk from a man wearing a hat, a starched white shirt with rolled-up sleeves and chewing on a cigar. He evidently was expecting me, and without a word took the examples of my photography work from me and silently examined them. After he finished, he looked up at me and said in a gruff voice, "Well, what do you want from me?" I thought it was pretty obvious and told him I wanted a job in his department. He said curtly, "We don't have any openings, try the News Room." Unceremoniously rejected, I slunk over to the large adjoining News Room. After a brief interview with one of the editors, I was offered and accepted a job there as a copy boy. I so very much wanted to be a staff photographer, but instead ended up in the News Room as a lowly copy boy, or as we called ourselves, "Lowly-of-Lowlies." Among the people with whom I worked there was Ben Bova, then an editorial clerk and fledgling science fiction writer, who went on to achieve fame as an author.

Over the ensuing months I fell into a routine of late work shifts, since the *Inquirer* was a morning newspaper. After running copy around all evening, I would meet Irv Lipkin and his friends for coffee and pie at Linton's restaurant.

Several of my fellow workers had waited for me to finish my shift at one a.m. and I met them downstairs with their car. We headed for the adjacent Press Bar, where we each downed a couple of Boilermakers. I, Seymour, the youngest of the group at age seventeen, was blasting off to experience my first drunk. Upon piling back into the car, I discovered a bottle of cheap sherry wine, a fifth of Southern Comfort whiskey, and beer, of which we all liberally imbibed. In no time I became a stumbling drunk and after cruising around for a while, we found ourselves in a seamy part of town at 8th and Race Avenues, where we spied a tattoo parlor. Someone suggested that they were tired of calling me Seymour and that I really ought to have a nickname. A contraction of my first name was decided upon and we entered the shop where I, a newly and totally inebriated Seymour Abraham Liebergot, ponied up fifty cents to have the tattoo artist permanently emblazon my new nickname, "Cy," on my upper right arm.

As dawn broke, my friends dropped me home at the street curb. Since I was incapable of standing, I was reduced to crawling across the sidewalk to the front door that resisted my fumbling attempts to unlock it. Finally successful, I continued my painful crawl up two flights of stairs to my bedroom where I gratefully passed out on my bed.

Three days later, after I had ceased throwing up, I realized that the tattoo WAS SPELLED WRONG! I also discovered that having a tattoo was socially unacceptable. Suddenly I was no better than "motorcycle trash."

I found that nicknames are given, not taken.

Andy Khinoy was one of the news editors who harbored some behavioral quirks. He would tirelessly edit newsprint and reporters' stories through the night, and occasionally, without raising his head, would toss a piece of copy into his wire OUT tray and call in a deep voice that I can only describe as mooing, "Booyy!" I, as a copy boy, a Lowlie-of-Lowlies, would run from my station at the Copy Desk to pick up the offering and hustle back to deliver it to the Head Copyreader, Mr. Chapeau, a short, erudite, gentle man, whose legs had long before betrayed him to a life with crutches. Regally, he sat in an elevated chair at the center of the Copyreaders' Desk and assigned copy to each of the eight harried copyreaders sitting around that horseshoe-shaped desk.

Andy also relished (no pun intended) sending a copy boy to the company cafeteria with weird sandwich orders, much to the scorn of the sandwich makers. He set me up for a lot of snickers one night when he sent me with an order for a sandwich of peanut butter and Lebanon baloney, with chopped black olives on cracked wheat toast. But he ate it. In fact, it became a regular order. For myself, I've never been tempted to re-create that concoction, for fear of the gastric consequences.

Each night I would visit the pressroom where the great three-story tall printing presses waited to be unleashed. I never lost my sense of excitement as I watched the pressmen, wearing their square signature hats cleverly fashioned from newsprint, prepare their beasts for the run. Once energized, a press turned agonizingly slowly, the winding sound increasing as the rollers picked up momentum. As it gathered speed the noise increased to such an intimidating point that I feared it had to fly apart, but it never did and I gathered up the first twenty-five copies to be quickly distributed to the editors and copyreaders for their never-ending search for errors.

— Chapter 24 —
1954: Decision Time

It was the fall of 1954, I had turned 18, and had spent sixteen months as a lowly copy boy. Although there was no doubt in my mind that I had more than mastered the limited duties of the position, there seemed to be no possibility of promotion. In fact, the next step up would have been to Editorial Clerk and the only additional duty would have been to write obituaries, which was never actually allowed. Consequently, I began to have serious doubts about my future in the newspaper business. My frustration was confirmed when a guy who had worked in the newspaper library (popularly known as the Morgue) for twenty-two years, was finally promoted to Copyreader. Whoopee! I imagined a ludicrous picture of myself still a copy boy after twenty-two years and I knew I had to make a decision about staying or leaving. Agitated, I walked over to the News Room Photo Editor (photo caption writer) and told him what was troubling me and that I was trying to decide whether to stay or to leave; possibly to join the Army. Without hesitation he raised his arms as if to exhort me and replied loudly, "Get out! ... Get out! ... Get out!!"

That afternoon I gave my two-week notice.

— Chapter 25 —
September, 1954: Army Enlistment

The day that I decided to enlist in the Army, Irv Lipkin and I were out walking and came to the US Army Recruitment kiosk located in a little patch of a park. I told Irv, "Wait here, I'll be right back," and I went inside and enlisted for a three-year stint. He yelled at me, saying that I must be crazy, but it was time to leave Philadelphia. Officially, I left on September 23, 1954.

A great day.

Junior Richardson, my best black friend cried when I told him that I was leaving for the Army. He said once you leave the ghetto, you never return.

He was correct.

The year before, I had tried to enlist in the Army when I turned seventeen. I knew even then that staying in Philadelphia was a trap. I hated my life and everything about it. My father would not give me permission. Why not? He apparently knew then that he had cancer. It was hardly a surprise, because after all, he was a five-pack a day cigarette smoker. He had been told that he had spots on his lungs and he had problems with swallowing. I suppose he knew that he was dying and simply wanted me to stick around. I was oblivious to all of this, so when I turned 18, I enlisted, this time without my father's blessing. I recall how sad he seemed at my news.

Sol died within a month of my departure. Phyllis told me he collapsed on the sidewalk, gasping for breath and asphyxiated from the esophageal cancer. A passing doctor tried to do an emergency tracheotomy with a pocketknife, but to no avail.

So, on October 13, 1954, a wasted life was finished at age 43.

Damn! I had just begun basic training at Camp Gordon, near the city of Augusta, Georgia, when I was called into the company commander's (CO) office to receive the news of my father's sudden passing. The local Red Cross helped get me back to Philadelphia so that I could attend my father's funeral. Everyone assumed that I wanted to go; they would have been shocked to learn that I hadn't the slightest desire. But I kept my feelings to myself, and passively accepted the help for the trip.

As I approached the open casket, all the anger that I had within me rose like a gorge, and I softly cursed the man for giving us such a crappy life. I now had my first debt to repay: my travel expenses that had been advanced by the American Red Cross. Also, I was "washed back" in Basic Training and had to start all over again.

Thanks, Dad.

— Chapter 26 —
Life After Sol

After Sol died, Josephine Oliveri was left with a lot of debt, no money, and three young children: Marvin 11, David 9, and Denise 3. After seven months of struggling against the tide, she finally capitulated and sought help. The final blow came when she was informed that she was not eligible for his Social Security benefits because she and Sol had not married. As far as the system was concerned she and the kids simply did not exist.

In May 1955, Marvin and David were placed in separate orphanages, and Denise into a foster home.

Eerily, history seemed bent on repeating itself.

Severely traumatized, Josephine retreated to Kingston in order to be with close relatives, and took up residence on, of all places, Josephine Street. She met and married Bud Youmans, a kind and gentle man, who allowed her to bring all the kids back after they had spent two long years in foster care. After a miserable time with Sol, Josephine finally experienced the good life, Bud gave her six very happy years until he passed away in September 1963, but at least he left her a small railroad pension. She died of heart failure on January 4, 1987, at the age of 66.

I live with the belief that Marvin, David and Denise resented me in that I had failed to maintain contact. Perhaps this autobiography will serve as a vehicle to reconnect with my past.

After Sol's funeral, I swore to myself that I would never return to Philly, but I had to do so briefly for my mother's funeral. Never having recovered from her nervous breakdown, Ida died of heart failure in 1968, at age 58. The Soloff family financial resources were limited, so Ida's body was brought to the cemetery in a cheap, wallpaper-covered pine coffin and, because such was

Josephine Oliveri, 1984

the hate of her family for Sol, she was buried in a part of Montefiore Cemetery separate from him.

How could any of us five offspring of Solomon Liebergot possibly mature into normal, stable adults? Let's briefly review his legacy:

Phyllis married Chuck Pulley in 1954 and produced three children: Philip, now a real estate developer, Steven, an ER Doctor, and Jacqueline, the youngest, a student.

Seymour (Sy) married three times and produced three children. He is now happily married, and has had a fruitful engineering career.

Phyllis Liebergot Pulley and husband Chuck Pulley, 2001

Niece Jacqueline Pulley, 1984

Nephew Dr. Stephen Pulley, 1998

Nephew Philip Pulley, 1998

suppose that genes don't always define our personalities because my father's two brothers, David and Milton, were such gentle and caring individuals. Could it be that some sort of genetic code came into play for the children in that all three brothers produced one son each (by a Jewish wife) and, even though there was absolutely no relationship to our fathers' vocations, encouragement in that direction by the fathers, or even mutual influence amongst the sons, the three sons ended up as engineers?

Marvin graduated from Kingston High School. He enlisted in the Air Force, but got out on a hardship discharge two years later in order to help his mother Josephine. He is now married and is a recruiter for a large securities broker.

David joined the Air Force, earned a college degree, rose to the rank of Lt. Colonel, and retired after 37 years in the service of his country. He is enjoying the fruits of a long marriage in the form of his five grown children and a number of grandchildren.

Denise has been happily married for 15 years.

Survivors, All!

1996: Half-brother Marvin Liebergot

Half-brother David Liebergot and half-sister Denise Liebergot Flanagan

— Chapter 27 —
1954: The Army Years – Basic Training

After returning to Camp Gordon via Fort Jackson, South Carolina, I was assigned to a new company to restart my eight weeks of basic training.

Before beginning basic training, recruits were administered many tests. One of them was the Officer Candidate Test (OCT), which was used to qualify individuals for Officer Candidate School (OCS). Apparently I did well enough on the OCT for the commanding officer to call me into his office and try to convince me to go to OCS. There were two things wrong with this: I'd have to re-enlist for six years, and I would have to serve in the Infantry or Artillery. I had opted for the Signal Corps at the time of my enlistment, because I had vague plans for a future that involved some sort of technology.

Another decision made.

Why did it seem that the Army shipped people from the North to posts in the South and vice versa?

Physically, basic training was not all that much of a challenge, as I was in excellent physical condition. Being basically a city boy, I had no experience in firing a weapon of any kind, so I was awed by the M-1 Garand rifle, the classic rifle from World War II. To become familiar with the weapon, we first fired on the "1,000-Inch Range" which was roughly 25 yards. I wasn't the only one who'd never held a rifle. I was incredulous when one of the 'cruits became so nervous about firing that he wet his pants; it was impossible to hide that dark stain when he stood. I was certainly nervous, but I didn't copy his performance. Firing the rifle

1954: Sy Liebergot beginning Basic Training

1954: Transient Barracks at Fort Jackson, South Carolina on the way to Camp Gordon Georgia

at the Big Range was a challenge, especially when it was raining. I recall lying prone in the mud created by the Georgia red clay, a poncho draped over me, trying to identify the target through my rain-splattered eyeglasses. Every time I fired, the bolt would fly back and splash water on my glasses, which I had to wipe with a handkerchief. When a cadre supervising the line poured oil on the bolt of my rifle, presumably in an effort to prevent rust, it caused my glasses to become splashed with water *and* oil which, when wiped, simply smeared the oil around making it virtually impossible to see the target. *Such help...*

Shorty McQuaig was a memorable character who deserved his nickname because he was closer to four feet tall than the minimum height requirement of five feet. How he got in, I couldn't figure; he was the shortest person in my company of 220 'cruits. Poor Shorty, he applied himself harder than anyone to master the required skills, epitomizing the phrase "try harder." His short stride kept him eternally out of step during marching drills; his head could clearly be seen bobbing up and down. Climbing up or over structures was nearly impossible for him. Usually a couple of us would provide the lift to help him. One of the training courses had a "Confidence Tower" constructed of four large wooden poles which were fifty feet tall and spaced apart to accommodate horizontal boards. In effect, they were very high ladders. However, the spacing between the rungs increased with height and, at the forty foot level, even a normal-height person was required to jump slightly to reach the final rung. Once at the top, the only way down was by a slide down a thick rope. Hence its name: Confidence Tower. For Shorty, this was an impossible task, so some of us would climb back down and literally hoist him up over the more widely spaced rungs. The cadre personnel never objected to us helping out; it certainly helped to enhance team spirit. Perhaps they just admired his determination. Hell, we all did.

Basic training was just that: learning basic military and survival skills, all laced with large measures of harassment. Eating the army chow that was served up during those eight weeks could probably be counted as survival training. I never knew where I was during night training, even when I had a map and compass, a failing I have retained to this day. However, I did become proficient with the M-1 Garand, with which I qualified for a Sharpshooter medal. I never qualified for another Army medal that I could wear on my uniform, not even the National Defense ribbon.

1954: Sy Liebergot finished Basic Training at Camp Gordon, Georgia

December 1954 brought the eight weeks of training to a finale. The final week began with a 16-mile march with a sixty-pound field pack (knapsack) to the bivouac area where we would dig our foxholes and erect our pup tents. When we reached the campsite that would be our home for the next week, the temperature was a balmy 82 degrees F, but Georgia demonstrated what a weather front could do to local conditions. Overnight, the temperature plummeted to 20 degrees F, and a couple of inches of snow fell. The freezing weather stayed with us for the entire week, creating somewhat realistic battlefield conditions, and to us eighteen-year-old 'cruits, it seemed like a fairly good re-creation. Was our bivouac to be a re-enactment of the Battle of the Bulge?

1954: Sy Liebergot at the start of Basic Training Week 8 at Camp Gordon, Georgia

On Graduation Day we marched in formation for the assembled officers. I'm sure most of them wondered to whom that single bobbing head belonged.

— Chapter 28 —
1955: Odyssey of Unexpected Destinations

Hooray! Basic Training was complete and my orders sent me to Fort Monmouth, New Jersey, which was located near Asbury Park and New York City. This was to be where I would get my Signal Corps technical training. I relished the idea of enrolling in substantial courses that would be useful later in civilian life, and meanwhile would see me stationed in Europe. Useful training and worldwide travel; that was *my* Army! ... I was sorely mistaken.

Eighteen Regular Army enlistees (RAs) and I were gathered in an empty room and informed that we would not be allowed to take any of the regular training courses for which we had applied. It was our (mis)fortune that the Army had created a new unit called the Army Weather Observers Corps (AWOC) and we were to form the core of this new group. It turned out to be the most abbreviated of all the courses; only six weeks in length. I later characterized it as spending six weeks in a revolving weather vane. The possibility of a posting in Europe was dangled as a carrot. The scheduled training passed quickly. After receiving an accelerated exposure to meteorological subjects and a bit of hands-on training with balloons and associated equipment, even generating hydrogen, we all graduated and were told of our next destination. It would not be Europe. It was to be Fort Huachuca, Arizona.

Hwa-what? Hwa-where? We couldn't even spell or correctly pronounce the name. We all ran to a map in order to locate this exotic-sounding destination, but to our complete dismay it appeared as if we were going to an utterly desolate place. *And in any case, why was the Army so interested in the weather in Arizona?*

1955: Beginning training at Fort Monmouth, New Jersey

— Chapter 29 —
Fort Huachuca: The Beginning of Hell

1955: Snow on Departure Day from Fort Monmouth, New Jersey for Fort Huachuca, Arizona

We departed Fort Monmouth in February, 1955, amid freezing temperatures and lots of snow on the ground. After a two-day train trip, dressed in our O.D.s, heavy winter uniforms, we disembarked in Benson, Arizona. Where in Hell was Benson, Arizona? Probably close to Hell, we thought, as the temperature was more than 90 degrees F, and we were sweltering in our uniforms. Benson was just a road and parallel rail junction with a cluster of buildings. We peered in both directions for our transportation to the new post, but saw nothing, so we found the sole soda shop, ordered up some sodas, and waited.

A military bus showed up within the hour, and we were told that Huachuca was fifty miles off. The only break in the monotonous scenery was

A dwelling in Benson, Arizona

when the bus whisked through Tombstone. *Tombstone?*

After about an hour and a half of climbing through the high desert we entered the main gate, but there still was little to see. We soon discovered that we were literally "in the middle of nowhere" at an altitude of about 4,600 feet in the desert at the base of the Huachuca Mountains.

Fort Huachuca was originally established in 1877 to fight Indians and outlaws. It had been home to the famous Black 10th Cavalry "Buffalo Soldiers" who arrived in 1913, served in Pershing's punitive expedition against Pancho Villa in 1916, and helped guard the U.S.-Mexican border until 1931. The post was declared surplus after World War II, but was reactivated during the Korean War by Army Engineers. In 1954, command of the post was passed to the Chief Signal Officer who found the setting and climate ideal for testing electronics and communications.

So, in February, 1955, we arrived at a barely reopened old Cavalry post that had become the home of the U.S. Army Signal Corps and Military Intelligence Center. It included the headquarters of the fledgling Army Weather Observers Corps to which eighteen other minimally trained Army weathermen and I belonged.

Although it became clear that Fort Huachuca was to be a Temporary Duty station, that didn't prevent the resident cadre from harassing us "newbies" with "temporary duty" assignments, none of which, of course, was weather-related. The M-1 Carbine (rifle) was the Signal Corps' official weapon, and so we were required to be proficient in its use. One day, we received word that we would be issued brand new carbines. The news created a bit of excitement (it didn't take much to get us excited). However, that enthusiasm was quickly dampened when we were each given a brand new carbine that was encased in a solid block of Cosmoline, an effective preservative that was virtually impossible to quickly clean off. We were then told that there would be an in-ranks inspection of the new weapons the next morning. We worked all afternoon and evening to clean the rifles, first by cutting away chunks of the stuff with a knife and then by applying a solvent, but we failed miserably. Of course, it was just the Army's way of using G.I.s to clean fifty carbines. *Very creative.*

Fortunately, after only six weeks at Fort Huachuca, all nineteen of us received orders to report to a new post, named Yuma Test Station, Arizona. *Yuma? Yu-What? Yu-Where?*

Well, after once more running to a map, we located Yuma on the Colorado River right at the border with Mexico.

Present entrance sign to Fort Huachuca, Arizona

— Chapter 30 —
No, This Is Hell – Yuma Test Station

I spent the remainder of my military career at Yuma Test Station and the experience had a profound influence on the direction of my life.

It was still 1955, the month of March, and our small group journeyed westward across the southern border of Arizona, through the outskirts of the town of Yuma, then thirty miles north, finally arriving at the gate of Yuma Test Station (YTS). Once again we were in "the middle of nowhere." I was at my final Army destination; assigned there until the summer of 1957 when my hitch would expire.

Although located on 2.5 million acres of desert, YTS was a small post, housing only 500 permanent personnel. It was a strictly technical post and I was gratified to find that there were air-conditioned working and living quarters, permanent KPs (Kitchen Police) in the Mess Hall and guard duty belonged exclusively to the MPs (Military Police).

Security was considered a serious matter since the post boasted a Research and Development (R&D) area three miles from the main post and a 75-mile long artillery range. The R&D area was the home of the Weather Station where my job involved knowledge of sensitive technical information and installations. A Top Secret clearance was mandatory.

In summer, the in-the-shade air temperature soared to a consistent high of 110 degrees Fahrenheit, sometimes topping 120 degrees. The "saving grace" was that the relative humidity ranged from only three to ten percent. Loose clothing and sun beanies (pith helmets) made up a mandatory uniform code.

And summer was nine months of the year.

Yuma was rated as the hottest place in the world for averaging more days per year with a peak temperature in excess of 110 degrees. Perhaps a bit of an exaggeration, but close.

Unlike Fort Huachuca and its near mile-high elevation, YTS was only 30 feet above sea level; the nearest elevated terrains were the Chocolate Mountains and the Imperial Sand Dunes, about fifty miles distant.

An anonymous soldier authored the following poetic assessment of duty at YTS:

Yuma Test Station Epitaph

Just below the River Colorado,
Yuma is the spot.
Where we are doomed to spend our time,

In the land that God forgot.

Here with the snakes and lizards,
Where a man is always blue,
Right smack in the middle of nowhere,
And a thousand miles from you.

We're supposed to be soldiers,
Helping to defend our land,
But what good are we doing,
In this desert heat and burning sand?

We're soldiers of the Sixth Army,
Earning very little pay,
Guarding people with thousands,
For two and a half per day.

Living with our memories, wait...
Waiting to see our gals,
Hoping that while we're gone,
They won't marry our pals.

No one knows we're living,
No one gives a damn,
At home we are forgotten,
We belong to Uncle Sam.

The time we spend in the Army,
The times that we have missed,
Boys, don't let the draft get you,
And for God's sake, don't enlist.

When we die and go to heaven,
To Saint Peter we will tell,
We soldiers at Yuma Test Station,
We have spent our time in HELL.

As the Army's hot weather testing station, the summer testing boosted the post's population to 2,500 personnel. Artillery personnel fired big guns to establish ballistic firing tables, the Quartermaster Corps marched troops around the desert to test tropical clothing and dug up food that had been buried the year before to determine if it was still edible, and finally, tanks and other armored vehicles were run on the dust course to sorely test the filtering systems. No doubt this served our military well during the "Desert Storm" Persian Gulf War.

The Weather Station was fully equipped and up and running, albeit in need of additional personnel ... that, of course, was us, "The New Nineteen." It was complete with surface observers, Rawinsonde (upper atmosphere balloons), field crew, repair shop, and two civilian meteorologists (I wondered what the Army had to do to get them to work in such a desolate spot). The station was important in that it provided weather data to our local artillery guys, the Yuma AM radio station with the oxymoronic call sign letters K-O-L-D, and the Yuma Air Force Base which was located on the outskirts of the town of Yuma.

I was one of four G.I.s assigned to the Rawinsonde Section. In fact, Air Force personnel conducted the Rawinsonde balloon launches. Upon learning of the limited training that we had received at Fort Monmouth, which they thought was a joke, they proceeded to give us OJT (On the Job Training). Teach they did, and learn we did. I loved the technical aspects involved in using real-time data recorders, adiabatic charts, slide rule and trigonometry.

1955: Rawinsonde computation room in YTS weather station.
L-R: Sy, Al Freberg, Tony Soteropolous

Little did I realize that I was on my way to an engineering career.

Army life settled into a routine centered on weather station activities. John Hruby and I formed one of the two teams that launched Rawinsonde balloons twice a day, seven days a week. Al Freberg and Tony "The Greek" Soteropolous comprised the other team. Those radio-telemetry balloons reached altitudes in excess of 100,000 feet, and sent information on their altitude, and the temperature, relative humidity, wind speed and wind direction. It was always exciting to launch those balloons from the roof platform of the station, aim the automatic tracking antenna called a GMD (Ground Monitoring Device) on it, and then run downstairs to watch the data come in on the

1955: Sy preparing to launch a rawinsonde balloon from the platform on the YTS weather station

data recorder. We had to speedily calculate and plot the downlinked information, and convert it into a block code to be teletyped to the air base and other parties. This took two hours. With two launches per day, we were kept pretty busy. Today, computers have replaced our hand-plotted adiabatic charts, crude circular calculators and slide rules.

Most post personnel were convinced that we weather station guys were a bit zany. We did our best to promote that opinion. There was the time, during windy conditions, that I "boloed" (wrapped) a Rawinsonde balloon train around the R&D area power lines, the train pulled the main power wires together, shorting out most of the area power as well as to the Main post. Another time, we launched a small pilot balloon, called a PIBAL, with a sign hanging from it inscribed with a silly statement. It had just enough helium in it to float at fifty feet above the ground. It stayed up all day, too high for anyone to retrieve it. Then there was further confirmation in the *Weather Words* weekly column I wrote for the post newspaper, in which I satirized the activities of the weather station personnel.

YTS The infamous hydrogen generator

Our supply sergeant failed to order sufficient helium bottles for our twice-a-day Rawinsonde and other balloon launches. So what did we do? We made hydrogen with a "hydrogen generator" that we had in the weather station. The Air Force had given up this practice due to the danger of an explosion, but not the Army. The Army was fearless: it ordered a couple of us Rawinsonde operators (one was me) to make the hydrogen. After donning rubber boots, rubber apron, rubber gloves and protective goggles, the recipe was simple: fill a strong gas cylinder with water, pour in two large boxes of aluminum chips, slip in three window-sash-size cylinders of sodium hydroxide (lye), and then rapidly screw on the tank cap and stand back to watch the pressure on its gauge quickly build to 2,200 psi.

The caustic aluminum hydroxide waste from this process was buried in the ground in a 4x4x6-foot deep hole that John Hruby and I had dug. Harassment? Let me count the ways...

During the third time that I generated hydrogen, an over-pressure burst disk blew out releasing a high-pressure jet of super-hot, corrosive aluminum hydroxide and explosive hydrogen. Since I was nineteen and indestructible, and not wanting to waste the hydrogen, I grabbed a replacement disk and a crescent wrench and proceeded to unscrew the retaining cap, replaced the disk and successfully screwed back the cap. All the while, the rubber gloves and apron were dissolving and the danger of explosion ever present. Having saved the tank of hard-won hydrogen, I was a hero. Our company commander was shocked to discover that we were even fooling around with hydrogen. The Air Force guys, of course, did not hesitate to tell us we were crazy. The hydrogen generator was taken out to the desert and "shot." (Not literally, however it was thrown on the trash heap.)

1956, YTS: John Hruby standing in the "necessary" 4x4x6-foot hole he and Sy dug to bury aluminum hydroxide waste from hydrogen generation

I dodged yet another bullet.

I succumbed to severe eyestrain from the close work plotting adiabatic charts with a sharp 9H pencil, and was allowed to move to the Field Crew. We had many field satellite weather stations scattered around the desert and they had to be serviced with fresh charts, ink and storage batteries.

I was teamed with Earl McCracken whose bulldog chin countenance reminded me of the movie version of an Irish New York cop. We had access to all areas on the huge reservation, irrespective of its security classification, if it had weather equipment that required servicing. One

1955: Sy visiting a field satellite weather station atop Luguna Mountain

of the restricted areas scattered around the desert was the Chemical Storage Area, a high fenced-off place that contained cylinders of nerve gas and other toxic products such as mustard gas stored under an open-air pavilion. Chemical experiments were still conducted at the R&D area where our weather station building was located. Mac pulled the truck up to the padlocked gate and I climbed onto the hood to observe the wind direction from our automatic weather station and to check to *see if the rabbits were still alive in their hutch next to the hundreds of cylinders of nerve gas.* If they

1957:YTS Jazz band L-R: Vic Sage - bass, Strawberry - conga, Glen Beckfield - piano, Sy - drums, Bill Burton - marimba (Leader)

were, then and only then would we go in to service our remote weather station. We were not issued protective clothing of any kind, so in stark contrast to more prudent attire for the potentially lethal environment, but in deference to the heat, I was outfitted in only my white skivvies, a pith helmet and army boots. I was quite the fashion plate. Overall, it was the ideal G.I. job, because no one ever knew where we were in the unmarked desert, and we came and went as we pleased.

    I loved music, especially modern jazz, and spent a lot of time in the Service Club playing ping-pong and listening to music. I heard a fellow playing piano in a side room and sat listening to him. After a time, he told me to set up the drum set and he set about teaching me what I needed to know so I could accompany him. Eventually this led to the formation of a jazz quintet that stayed together for more than a year, and we played for beer and eats. I recall vividly the terror that I felt when Bill Burton, the leader, told me, "Take it," for my first drum solo.

More exposure to live music.

The most eccentric character in our barracks was Isaac Harter III, from Beaver, Pennsylvania. Isaac was a scion of a rich family who had been wasting his life tinkering and was drafted into the Army after being thrown out of several Ivy League colleges. He frequently talked in his sleep and on occasion entertained us by fighting sea battles while in that state. At some point after falling asleep, he would sit upright in his bunk and narrate his involvement in a great sailing ship battle. "More powder, Mr. Jones," he would shout, "we're ten yards short!" Then there would be a period of silence, followed by, "Egads, they've made Stratton walk the plank!" (Stratton was our barracks sergeant.) We sat near his bunk, rapt, wondering how the story would develop. Isaac remembered none of this after waking, and denied it even when I replayed his sleeping adventures from a tape recording I had made.

1957: Now a drummer in the Service Club on Yuma Test Station

The end of a month was a tough time for an enlisted soldier. We were paid once a month and tried unsuccessfully to meter our expenses to last; a pack of Lucky Strike cigarettes was considered barter. It was quite ludicrous to see us playing poker with the pot consisting of a pile of cigarettes and one of us raising a player "two Luckies." None of us were very successful at rationing our meager pay, but the NCO Club did its part by offering spaghetti and "cheap drinks" nights during the last week of each month. On one of the cheap nights, when drinks were 50 cents, Sal Randazzo walked over to the patio table where I was sitting with a sergeant whom I knew, threw two quarters down on the table and announced that it was all the money he had remaining. He then bet he could drink a Zombie (a 12-ounce mixed drink made of various rums, liqueurs, and fruit juice) through the two little bar straws in less than thirty seconds and thereby win another drink. The sergeant said, "no way," and took the bet. An additional condition was that Sal had to walk the distance to the bar to get his prize. To our astonishment, Sal sucked the first one down in 20 seconds. Sarge renewed the bet. I timed the next seven Zombies at 15-20 seconds each, and each time the Sarge said, "no way." Once again, Sal rose from the table, passed through the large crowd that had gathered to watch, and marched into the Club for number eight. Moments passed, then more, and just when we were thinking of check-

1957: Sy as a diver on the YTS swimming team

ing on him, out he came, his drink held firmly in his hand, his legs making walking motions, but his feet were not touching the ground since he was held up by his arms between two large grinning bartenders. With great ceremony they gently sat him in his chair, his hand still clutching his Zombie, but his eyes were unseeing and unblinking; Sal had left us. Talk about becoming zombie-like. He was sick for days.

The swimming pool on post was of dubious utility when the temperature in the shade was 115 degrees and the water temperature was almost 100 degrees; we were provided brief relief when we jumped into the pool.

Sgt. Orick decided that YTS needed a swim team, and since I was a former gymnast, I was assigned to become a springboard diver. However, after several coaching sessions, I still could not perform a simple front dive. The coach declared that I would never be a diver. Never say never to me! I checked out a book on diving from the post library and taught myself in private. One of my greatest pleasures was showing the coach up at my first swim meet in which he was a diving judge.

The Command of the Weather Station was not a popular assignment. Captain Hewitt was our third commanding officer. He was a short man who drove the biggest Pontiac he could find with a built-up driver's seat. He was a Regular Army officer with an abiding dislike for civilians, whom he had annoyed in his last command in Canada, where cold weather equipment testing took place, and his transfer to YTS was a punishment. He had implicit trust in his noncoms and turned a deaf ear to our complaints about First Sergeant Middlebrook, whose excessive harassment and pettiness we deemed to be unwarranted because the weather station was a working, highly technical installation and not some basic training unit. Unbeknownst to anyone, however, one of the surface observer fellows had been steadily complaining to his mother about Sgt. Middlebrook's treatment of him, and the rest of us lowly privates. Like a good mother, she wrote letters to President Eisenhower, passing along the complaints and of course, expecting results. Two Army captains from the Inspector General's Office in Washington, D.C. showed up unannounced at Captain Hewitt's door to conduct an investigation. The CO was floored, but he had no option but to let the officers investigate. They determined that all the complaints against Sgt. Middlebrook were well-founded and took measures to force the 23-year veteran out of the service. The last time I saw him, he was driving a taxicab in Yuma.

Yes, there is justice.

As 1957 dawned, I was approaching my twenty-first birthday, only nine months remained until I would be discharged, and I hadn't yet made any decisions with regard to what I would do with my life. I still lacked the confidence that I could do college-level academic work. None of my buddies had any plans for their lives after getting out of the service, but privately I felt uneasy about my future. After all, I had joined the Army to "take a break," get some technical training, and figure out where I was going with my life, hadn't I? The answer came unexpectedly. I was visiting the barracks of the Ordnance (artillery) outfit to discuss hi-fi electronics with some of the technicians whom I knew. Most of them were college students when they were drafted into the Army; the draft was still in force during those post-Korean War years. At some point, it dawned on me that these guys were no smarter than I was, yet they had been attending college as if it were no big deal (maybe that's why they were drafted), so why couldn't I? The feeling grew stronger over the next few days, and finally I knew that I could do it. I decided to move to Los Angeles, California and attend college for a degree in Electrical Engineering. I had protected my eligibility for the G.I. Bill, which included education benefits, by enlisting in the Army four months before the Korean Bill PL 550 expired.

I sometimes wonder how these decision points just seem to happen.

Yuma Test Station was an equal opportunity "origination point"; that is, it was located close to nowhere. It was 30 miles to Yuma, 60 miles to San Luis, Mexico (and all its delights), 205 miles to San Diego, 275 miles to Phoenix, and 305 miles to Los Angeles; I had visited them all. In late 1956, Los Angeles became my favorite escape, since there was always somebody with a car that was going there on the weekends and would carry passengers round-trip for five dollars. I began going every other weekend with a couple of buddies who had friends and relatives in LA. The trip to Los Angeles was a boring 6-hour drive through California desert scenery that included the Imperial Sand Dunes, quickly through Calipatria, past the Salton Sea, Palm Springs and punctuated by the pungent odor from the Coachella Valley (cattle) feed yards. Though the surface of the Salton Sea was 227 feet below sea level, the small town of Calipatria, at 184 feet below sea level, was the lowest community in the United States. In fact, so conscious of its geographic "lowness" were the townspeople, that they erected a 184-foot flagpole to allow the America flag to fly at sea level! The outdoor sign of a local eatery boasted of the "lowest" hamburger prices in the country.

It was on one of these trips that I met Deanna Cohen, with whom a relationship quickly developed. As the months dwindled toward the end of my Army active service, marriage became a real prospect. My close friends strongly counseled me against marriage, but I was emotionally unprepared to make such a major change in my life alone, so I ignored their advice.

I was honorably discharged on July 19, 1957, after serving two years, nine months, and twenty-nine days active duty in the U.S. Army.

— Chapter 31 —
A New Life: Los Angeles, California

With my discharge orders in hand I traveled to Fort Bliss, Texas where I was processed out of the United States Army with sufficient funds to return to Philadelphia, whence I had enlisted. However, I was headed west to Los Angeles, not east and I hitched a ride with a soldier also heading to LA. After we drove the distance non-stop, he dropped me off at the freeway exit to West Covina, California, where Deanna lived and continued on his way. I found a telephone and called Deanna to pick me up.

Post-army life began with a bang. In short order, I had to make wedding plans, find an apartment for us to live in, buy a used car, find a job, and enroll in college.

Deanna Cohen and I were married on August 4, 1957, which honored the birthday of my sister Phyllis who was unable to attend. We purchased a green 1953 Plymouth, and rented a one-room efficiency apartment with a Murphy bed.

My new wife and I thoroughly discussed and agreed to a plan for our life ahead. She was to continue to work as a dental assistant; I would find full- and part-time work and carry a full engineering school course load. We would be able to get by financially with the additional stipend from the G.I. Bill. We agreed that there could be absolutely no children, at least not until after I graduated with my degree.

The Plan For the Future was all so perfect, or so it seemed.

I found my first job as a shoe salesman at a ladies' shoe chain store called Leed's. It was a "fast," high sales volume store on Hollywood Boulevard, a half-block from Vine Street. I would keep this job for five years. Closing a sale quickly was paramount to earnings, which were based strictly on sales commission. Many times, a customer would become confused and indecisive with regard to whether she should match her handbag or her dress, and once she had decided, there was the question of how many pairs of shoes she should buy, and so on. One of the techniques that we salesmen employed was "creating a distraction." On one occasion, Paul Mardarocian, a fellow salesman, had a particularly slow-to-decide lady and called me over to chat. He began, "How was the gig last night with Stan Kenton?" Soon I was weaving a tale that I was a part-time sax player who would be called to fill in with various bands when they visited town. When the dialogue got to the point where "I had some of Kenton's band fall over to my pad for a party," Paul picked up the pair of shoes that she obviously wanted saying, "This way, ma'am," and quickly moved toward the cash register. Incredibly, as if in a trance, she rose from her chair and followed him without a word of protest; she was in the hands of a master. The ploy worked nearly all the time. Sometimes one of the many small earthquakes that shook the building provided the necessary distraction.

Soon after getting married, I hustled over to Los Angeles City Junior College (LACC) to enroll in time for September classes. The Administration Building was classic old-style architecture, sporting a capitol-like dome. As I walked toward the building I wondered, where had I seen it before? It dawned on me that it was the building shown in the opening scene of the TV series *Science Fiction Theater*, that ran from 1955-1957. While the LACC Administration Building was shown on-screen, host Truman Bradley intoned, "On the busy campus of a large metropolitan university, …" and the episode was on its way. I loved that show.

LACC enjoyed historic status because its campus was the early location of two well-known California colleges: UCLA and California State University at Los Angeles (CSULA). I would attend CSULA during my second two years of study.

A requisite for matriculation in LACC was a battery of aptitude tests. A counselor reviewed the results with me, and suggested that I should pursue a career as an English teacher, not engineering. She told me the best I could expect was being a "C" grade engineering student.

> *Another "you can't," challenge. It turned out, that "you can't" was one of my hot buttons; I didn't like being told that I couldn't do something.*

— Chapter 32 —
Family and College

Our first child, Shelli Lyn Liebergot, was conceived in the month of November 1957. When Deanna confirmed that she had indeed become pregnant, I was in shock; I saw all my plans for the future crumbling. Shelli was born in the Cedars of Lebanon Hospital on July 18, 1958, after a difficult 26-hours of labor. Deanna became pregnant again in July, one year later, and our twin boys, Mark Daniel and Scot Alan, were born April 16, 1960; a consequence of the failure of the hospital clinic to advise Deanna to refit her diaphragm. The twins were born six weeks premature and required an additional two months of care in the hospital. Fortunately, we qualified for charity medical treatment. So, in my sophomore year of college, I found myself carrying a full engineering course load, and working full- and part-time to support three children in diapers and a wife.

So much for the Grand Plan.

Soon after the birth of the twins, I sought a medical remedy to stem the profusion of children – a vasectomy. In the 1950s, doctors were reluctant to perform vasectomies. In fact, such was their reluctance that some had a scoring system in which the applicant's age was multiplied by his number of children and only if that result exceeded an arbitrary number would the simple operation be granted. The first doctor I visited applied that system to my request, and refused. I asked him if at age 24, would I have to father two or three more children to meet his 'standard'? No answer. However, I was later successful in my quest for a sympathetic doctor. His name? Dr. D.O. Dickey. That was his real name!

Thank heavens for the Korean War G.I. Bill and the California State Education system, or I would never have been able to afford college – even though junior college tuition was only $6.50 per semester and Cal State was $47 per semester. The G.I. Bill paid the basic tuition for 36 months (4 years) and, since I was married with three kids, a monthly stipend of $160. Including my part time work, we were living on a monthly income of $225, and so were barely scraping by.

Adjustment to my altered life was difficult. I was eminently unprepared to deal with children, but I had no option but to learn. I changed diapers and baby-sat when needed, but I was unable to perform the most important duty of being a loving parent because there was time only for the perfunctory things and besides, thanks to my own upbringing, I'd had no experience with effective parenting myself. The rest of my time was taken up with school and work. This put a real stress on my marriage. Deanna always deferred the disciplining of the kids to me. To my small credit, apart from delivering a few swats to a bottom with my hand when I believed it necessary, I never beat my kids. I was determined not to follow Sol's precedent in that regard.

Every day was filled with classrooms, my shoe salesman job and never ending homework. When could I sleep? I could afford only three hours sleep a night. Nevertheless, it never crossed my mind to abandon my goal of an engineering degree. I barely noticed that the Russian Sputnik was successfully launched into orbit on October 4, 1957.

One day on campus, I suddenly doubled over with horrible chest pains and soon thereafter added heart palpitations. I began to suffer one symptom or the other half a dozen times a day, and sometimes both simultaneously. With my breathing restricted and my heart feeling as if it was running away, I became frightened and sought the help of a neighborhood doctor. After examining me and listening to a recitation of my daily schedule, he told me that I was suffering from a psychosomatic reaction to stress and he gave me a small bottle of Deprol, a tranquilizer, with instructions to take one at noon each day. When I pointed out that I could only give up three hours per day for sleep, he gave me a bottle of Dexedrine (dextrose amphetamine sulfate) and prescribed one pill first thing, each morning. So there I was, taking an upper in the morning and a tranq in the afternoon. After several months of this "treatment," the chest pains and palpitations were virtually eliminated, but I continued taking the Dexies every morning for a couple of years until my schedule lightened. Happily, I did not become addicted.

Five semesters of class work allowed me to graduate from LACC and move on to CSULA to complete the upper division classes for my degree. I was unable to form many friendships, even with my study partners, because I just didn't have the time. The sad thing was that because I was totally focused on earning my engineering degree, I never enjoyed the college experience.

The last two years were a blur of engineering class work and studies, with little special to recall. Engineering lab reports were especially arduous, and so time consuming. We were given one hour of semester credit for three hours of lab class time each week, in order to prepare eight lab reports that took forty hours each to prepare. Larry Canin, my study partner and best friend, and I solved this problem in typical student fashion: We forewent sleep. We began work on a lab report early on Saturday morning and worked straight through to Monday morning, skipping sleep. This was made possible by sharing my Dexies with Larry. Sometimes this was not without consequence. As I was driving to school one Monday morning after having taken two Dexedrine pills the previous day, I suddenly realized that I was 'watching' myself drive the freeway from the back seat, up near the ceiling of my 1953 Plymouth. The drug-induced out-of-body experience was very scary.

As I neared the end of my academic studies, the regional manager of the shoe store chain for which I worked approached me with a job offer of an assistant manager's position in one year and my own store (as manager) in three years. I felt insulted, for he obviously had no idea of the importance of an engineering degree and my single-minded dedication to this goal. I had no intention of making the retail shoe trade a career, and held my tongue because I still needed the job which, after four years, I was more than ready to leave. My opportunity came when Larry, who was a semester ahead of me, had a job in the Space & Information Systems Division (S&ID) of North

1962: NAA College engineering Unit L-R: Phil Fagan and Larry Canin

American Aviation in something called the College Engineering Unit, and advised me to apply for a job there.

Located in Downey, California, North American Aviation was a famous airplane manufacturer, established in 1928, that made its mark during World War II when the company produced, among other famous airplanes, 9,498 B-25 bombers and 15,586 P-51 Mustang fighters.

On June 12, 1961, I applied for a job at NAA and was hired into the S&ID College Engineering Unit and entered the real world of engineering. I was one of 135 engineering students given an opportunity to apply our classroom-gained knowledge to practical, aerospace engineering. It was valuable on-the-job training. The job was part-time, in the evening, which was perfect for a full-time student with daytime classes. A year later, the group grew to 160 students and I became the Assistant Supervisor to Phil Fagan, supervisor and also a college student. Phil and I became friends during my two-year stay in the Unit and he was always ready to offer advice and guidance. When I doubted my abilities, Phil would tell me, "You're an engineer, and you can learn anything!" I found opportunities in later years, to repeat these words to other engineering-bound students.

Thanks, Phil.

Unfortunately, I was hired at a slack time. The company seemed to be in a holding pattern, awaiting its next contract. I remember wandering around the empty hallways of the cavernous main building wondering where our work would come from. I needn't have been concerned. Though I was fairly oblivious of our space efforts, Project Mercury was already underway in 1961, with chimpanzees Ham and Enos paving the way for the rest of us primates, and culminating on May 5, 1961 in Alan Shepard's historic sub-orbital flight in his *Freedom 7* capsule. A few months later, President John F. Kennedy publicly committed the nation to "landing a man on the Moon" before the decade was out. In September 1961, our division won a contract to produce the S-II, the powerful second stage of the behemoth Saturn V lunar launch rocket; in November, NASA chose the division to develop and build the Apollo Command and Service Modules, effectively making S&ID NASA's prime contractor in the lunar program. Less than eight years later *Apollo 11* landed on the Moon.

Talk about being in tall cotton.

I was inspired by the words from Kennedy's speech on September 12, 1962:

> *"We choose to go to the Moon. We choose to go to the Moon in this decade and do the other things, not because they are easy, but because they are hard, because that goal will serve to organize and measure the best of our energies and skills, because that challenge is one that we are willing to accept, one we are unwilling to postpone, and one which we intend to win, and the others, too."*

A JFK tape transcript of a meeting to discuss Supplemental (budget) Appropriations for NASA was released in 2002. The meeting took place on November 21, 1962 and contained an exchange between President John F. Kennedy and James Webb, then NASA Administrator, during which Webb told Kennedy that "he didn't feel a Moon landing should be NASA's top priority." Kennedy disagreed saying in part, "Everything we do ought to really be tied into getting on the Moon before the Russians ... otherwise, we shouldn't be spending this kind of money because I'm not that interested in space."

So much for the romance of space exploration; this was only a cold war tactic.

A close friend observed that the Apollo Project was similar to the Manhattan Project in that the waters were uncharted and the goal had to be accomplished in a short time frame. We were essentially at war both times; one hot, the other cold.

As 1962 dawned, I needed only another eighteen school hours to graduate, so it was to be my final year of college. Needing more earnings, I accepted a full-time position with an engineer classification and planned to finish my degree in two semesters at night. I hadn't realized the challenge that three electrical engineering courses at night would present. I finished the first semester with great difficulty. Although only nine semester hours remained, the toll taken by family, work, and school finally got to me, and since I was already rated as an engineer, I seriously considered giving up the chase for my engineering degree, but I dug deep and found the reserves for the last three elective courses and finally achieved my long-time goal of an engineering degree in January 1963. However, it was such an anticlimax, that I didn't attend the graduation ceremony.

Part Two

Mission Control

— Chapter 33 —
Shaky Start

With the space program hardware being developed and built at the Downey facility, my future seemed secure, so Deanna and I felt brave enough to take on our first mortgage, a G.I. Bill loan, and purchase a typical Southern California tract house in the town of La Mirada, 10 miles east of the Downey facility. The tract had been created out of an olive tree ranch; every house had a stucco-over-chicken-wire exterior and had an ugly olive tree planted in front. It was a blue-collar neighborhood. When it became known that I was an engineer, we were pretty much shunned, even our kids.

As the pace of the program increased, hundreds of engineers and support staff were hired to fill the old buildings and the temporary trailer buildings that were used while the new engineering Building 6 was being constructed. I never tired of going to the huge manufacturing area that was now filled with the Command and Service Modules (CSMs) at different stages of completion, and wandering among them observing the new manufacturing techniques, such as the introduction of welding aluminum in tents of helium and tape-fed computer-driven tooling machines. I was interested in everything.

Little did I realize that this was a step along the road to becoming a flight controller.

Where were you when John F. Kennedy was assassinated? The question is asked so often that I must also give my answer. The date was November 22, 1963. I recall so clearly talking to some engineers in one of the temporary trailer buildings when the unbelievable news began spreading: Kennedy, the 35th President of the United States, had been shot to death in Dallas, Texas. Lyndon Baines Johnson was quickly sworn in as his successor. What would happen to our fledgling lunar space effort? Would Johnson honor JFK's memory by continuing his program to land a man on the Moon in this decade? I decided not to entertain gloomy thoughts, and continued with my job as if it would never end.

Ah, youth and inexperience; the great insulator.

Soon thereafter, the company issued a call for Human Factors Engineers. It wanted people with an undergraduate degree in engineering, math, or physics and a Master's in Psychology. I was sorely tempted and began studying the requisite undergraduate psychology courses in preparation for a Masters. Dr. Gordon Wells'

Crew Systems group, comprised by psychologists one and all, had the responsibility of designing the interior of the Command Module. Specifically, they were to figure out where the switches and gauges would be best placed and conduct "reach" studies. Pretty important stuff, I thought, and I was gearing myself toward that field of endeavor. That all changed when Jim Ward, an engineer chum who worked in Crew Systems, told me that even after a couple of months of study some of the psychologists were still undecided on the best locations for switches and gauges on the Main Display Console that would be located directly in front of the three astronauts. One day, disagreements arose, got worse by the minute, and finally reached a crescendo of screaming punctuated by statements about who had published more technical papers. The task was taken away from the psychologists and assigned to Engineering, which completed the task in less than two weeks. I decided that while the psychologists controlled the field that I would shelve my desire to become a Human Factors Engineer.

In 1964, I got wind of a new group, called the Flight Operations Support Group, which was formed to provide CSM systems information to NASA Flight Operations at the Manned Spacecraft Center in Houston, Texas. I had no intention of moving to Houston, but the idea of being close to flight operations interested me and I arranged an interview with the supervisor, Charlie Ackerman. With Charlie's endorsement, I was transferred into the new group. Before the main body of the infant Flight Operations Support group moved to Houston, he promoted me to Lead Engineer and assigned two members of the group to me. "Moose" Morris and Earl Wylen, who were from Florida, resented being left behind and planned to return as soon as an opportunity presented itself. My new job was not to be easy.

A group of twenty engineers from San Diego who hired on with North American (when all the aerospace companies were staffing up for the Apollo program) but were very cautious in that they continued to live in San Diego, pooled their resources and purchased two passenger vans so as to comfortably commute the 250-mile round trip to Downey each day. After nearly a year of this grueling ritual, they all decided that the program was stable, sold their homes in San Diego, and moved their families to the Los Angeles area. Sadly, soon after that, S&ID had to make its first major layoff, and each of the ex-San Diegans were included in the 3,500 employees who marched out through the front gate over a 6-month period.

Such were, and still are, the vagaries of being an aerospace engineer.

— Chapter 34 —
July, 1964: Houston, Texas

Eighteen months after joining the Flight Operations Support Group, I became bored answering questions from the main group in Houston, as well as frustrated with having to direct two subordinates who wanted to be elsewhere. Combining that with the reality that my marriage was on shaky ground, I requested to be transferred to Houston in the hope that a change in scenery would help.

At midnight one day in early July, 1964, I sat behind the wheel of my 1962 red Volkswagen Beetle, with all of its 40 horsepower, and set out on the 1,147-mile trip to Houston. Sixteen hours later, having reached El Paso, Texas, I was feeling pretty good, until I realized that I was only half way to my final destination. It finally dawned on me just how immense the State of Texas is.

The next morning I made a planned side trip to Goodfellow AFB in San Angelo to visit my half-brother David, whom I hadn't seen since I left Philly in 1954 – ten years. I found him at pool side where he greeted me coolly. As he saw it, I had "deserted" him when I left for the Army. We had dinner that evening at a local restaurant and made some headway at reconciliation (I haven't met Marvin or Denise again, although Marvin, David and I have communicated as a result of this book). Early the next morning, I resumed my epic drive to Houston, arriving on Saturday, July 4th.

Houston was pretty much a hick town, compared to LA whence I came. The July heat and humidity were oppressive to me in my tiny un-air-conditioned Beetle. There was an electronic billboard on the roadside heading into town that counted the population of Houston: the count was 800,000. Now? More than 2 million.

On my way to apartment hunting, I stopped by for a brief visit at an Oshmans' warehouse, which was the temporary quarters for the NASA Flight Operations personnel. NASA was still in the process of establishing itself in Houston, and it was spread across the city in a variety of temporary quarters wherever sufficient office space could be found, such as the Oshmans warehouse and the Houston Petroleum Center office complex on the Gulf Freeway. The Flight Ops personnel were to report for work on Monday at the not-quite finished Manned Spacecraft Center (MSC), allowing me to report there directly.

In the early 1960s traveling anywhere around the Houston area was slow, because the roadway infrastructure was in its developmental infancy. The Loop 610 Freeway, which would eventually circumnavigate the city proper, and the Pierce Elevated, which would lead the Gulf Freeway through city center, had been barely started, and because real work on the Gulf Freeway hadn't yet begun the interchange at the Gulfgate shopping mall was a grassy mound of dirt. And futhermore, Farm-to-

Market road FM 528, which was only a small two-lane blacktop, had to carry the newly increased volume of traffic through tiny Webster to the cow pasture that was rapidly being transformed into the MSC campus. The existence of the space facility would soon give impetus to the creation of a new community named Clear Lake City, which was located near an inlet of Galveston Bay that was neither near a lake nor clear!

1962: MSC construction site

Deanna and the children flew down after I had found a furnished apartment and we set up temporary living for a few months. Daughter Shelli was six years old and the twins were four, so we had time to find a house in which to restart our lives, Texas-style.

We had a house built in Dickinson, 10 miles south of MSC. It was a new subdivision, called Bayou Chantilly, on the southern banks of Dickinson Bayou, and we settled in to what promised to be the normal life of a typical nuclear family. The house was located on a cul-de-sac and situated on a large pie-shaped lot. Soon, weekends were taken up with mowing the lawn, house repairs, and maintaining a Beetle that did not live up to its reputation for being reliable. In time, the kids enrolled in the local school and Deanna and I struggled to understand the New Math with which Shelli had to contend.

I made a couple of significant decisions that would prove to be beneficial, both physically and emotionally:

As soon as I graduated from high school, I succumbed to peer pressure and choked my way through my first cigarette; after all, my (sports) body was no longer my temple and I was invulnerable, wasn't I? Over the next ten years, I worked my way easily up to two packs a day without giving a thought to the cause of Sol's early demise. At age 28, in addition to my smoker's croupy cough, I contracted a cold that infected both ears. When it resisted standard treatment, Dr. Schnake placed me in the local hospital for three days of "intense treatment" *(Translation: multiple penicillin shots to the rump, administered by a sadistic nurse).* On the office visit following my discharge from the hospital, Dr. Schnake pronounced my ears clear. Since I still had the croupy cough, I inquired, "How come I don't feel much better than I did 100 dollars ago?" The good doctor recommended that I give up cigarettes. When I got home, I threw the remaining half-carton of Lucky Strikes into the trash and quit "cold turkey," then and there. One week later, accompanied by withdrawal symptoms, my lungs cleared up and the croupy cough disappeared, never to return.

Thank you, Dr. Schnake; another bullet dodged.

I always felt the need for music. A shrink might observe that it probably has something to do with suppressed emotions. After leaving the army and my drumming experience in the jazz group, I toyed with the idea of taking drum lessons. I wisely decided that it involved too much equipment and too much noise. Charlie Byrd was a musical hero of mine, so I took up classical guitar study in order to learn to play like him. At least that was the goal. I discovered Richard Mannix in Clear Lake City and he became my teacher and, after a while, my beer drinking buddy. He had a small mixed-breed dog named Bonnie who was addicted to cigarette butts and would perform tricks in return. Richard had a quirk in that he preferred to drink only beers of dubious reputation (at least they were not to my taste). On one occassion, he and Bonnie took a vacation tour around some of the states in the southwestern U.S. during which he collected memorable beers such as A-1, Lucky Lager, Shiner, Falstaff, Lone Star, and others and gave them to me as a present. *I drank them all.*

1973: Classical guitar teacher Richard Mannix

I studied with Richard for five years, but I quit when I failed to become adept at sight-reading classical guitar music. So I couldn't play like Charlie Byrd. Who could? I continued to play for myself, and put the lovely pieces that I had learned to good measure on my lady friends.

Live music was in my life again.

— Chapter 35 —
Flight Control: Beginning

As a Command and Service Module (CSM) spacecraft Sequential Systems specialist in the Flight Control Division, I provided NASA with current operating information of that spacecraft's many systems. My office life was filled with phone calls to the engineers in the Downey home plant to discuss the latest design changes, creating schematics for troubleshooting, writing crew procedures both for normal system operation and for dealing with malfunctions, and decision strategies for pre-planned mission flight rules.

The Gemini Program, in which I had not been involved, was drawing to a close in 1966 and experienced flight operations people were phasing over to the infant Apollo Program. The first mission was to be (Apollo-Saturn) AS-201, a 37-minute downrange rocket lob topped with an unmanned CSM that was automatically directed by a tape sequencer normally used on an Agena rocket stage. Agena was developed by the Air Force as an upper stage for the Thor and Atlas rockets. It had been used to launch the *Ranger* Moon probes and the earlier *Mariner* Venus and Mars probes. NASA used Agena for the Gemini project as a target for docking and rendezvous exercises.

The short flight successfully fired the Service Propulsion System (SPS) for the first time, and tested the sequential system that was to control the Command Module's separation, attitude control, and parachute sequencing and landing systems. We ran 144 training simulations of this relatively simple mission. As a North American Aviation employee, I supported the effort in a "back room" advisory capacity as a Sequential Systems flight controller. Upon returning after missing a week of sims through illness, I asked Mort Silver, who sat at the console next to mine, if EECOM Dave Pendley, in the "front room," had called back to me for information. Mort's reply was "not once." I began to understand the real role of a contractor employee; it was definitely not as a real-time flight controller. It was time to make a decision, because if I wanted to be in on the real action it would have to be as a NASA employee.

A year and a half after moving to Houston, I switched over to NASA and began my flight controller career in the Manned Spacecraft Center's Mission Operation Control Room (MOCR).

The work was so engaging that I found myself neglecting my family ... again.

— Chapter 36 —
The Singin' Wheel

The world of a flight controller revolved around Mission Control, The Singin' Wheel, and home, in that order.

Webster, Texas was reluctantly thrust into the future with the establishment of the core of the nation's manned space program just two miles to the east. Webster residents resented the huge intrusion of traffic through the very center of their sleepy bedroom community of 2,000 souls. This was evidenced by the zero-tolerance policy of the Webster Police shown to any non-residents. They were better known as the "Webster Gestapo," and stories of their mistreatment of people abounded. Examples: a "hippy" teenager with blonde hair in a long ponytail was riding his bicycle through the town late one evening and was unceremoniously thrown in jail for an unknown reason; a young mother, her infant, and elderly grandmother made the mistake of exceeding the speed limit of 35 mph whereupon they were pulled over by the Gestapo. The young mother was taken away to jail, and the infant and grandmother were left in the car by the roadside throughout the night.

2002: Singin' Wheel

Webster also was home of The Singin' Wheel restaurant and bar, also known as "The Wheel" or "The Red Barn" on account of its appearance, which became the official "watering hole" and second home of the flight controllers. It was owned by Audrey Long, a local realtor, and managed by Nelson Bland, whose personality and appearance were true to his surname.

Inside, the floor sloped unevenly and the wooden tables and chairs were sparse. In contrast, one entire wall was taken up by an immense antique bar of dark wood behind which was a large ornate mirror that dominated the wall. Three one-gallon jars on the bar advertised pickled pigs' knuckles in one, pickled purple hard-boiled eggs in the second, and pickled polish sausage in the third. Their presence was designed, of course, to make you very thirsty when eaten. There was a small building outside that housed a large smoker where the "cooking ladies" made passable

BBQ beef and wonderful ribs. Of course, the Singin' Wheel Chili Cheeseburger was not to be avoided. Copious pitchers of beer were always consumed.

It was a great place to eat.

It seemed that drinking beer was a flight controller's pastime. When a group of us convened at The Wheel, we'd start a pile of dollar bills in the center of a table and order pitchers of beer. As additional people showed up, they would toss dollars in and keep the kitty going.

It was a great place to drink beer.

The Wheel was a place for us to play. A downstairs back room became the musical home of the "Backroom Swingers" with Johnny Ferry and Larry Canin on guitar, me on a small snare drum, and Hershel Perkins on "chair back," which he attempted to strum, in the process ripping much of the skin from his fingers; John Hatcher showed up on occasion with a "tub bass" and we "swung." We all tried to sing; but Ferry was the only legit singer.

It was a great place to have fun.

The Wheel had a small club upstairs, complete with a bar and a tiny dance floor. Audrey Long remodeled the club with antique furniture and created an intimate ambiance. There was an old upright piano and most nights you could find Thelma playing the piano and Dinky singing. They were a dichotomy of somatotypes: Thelma was a matronly, heavyset woman in her fifties, while Dinky was a trim, pretty blonde whom we all tried in vain to hustle. I would sit in occasionally with my snare drum. Needless to say, on many a night we returned to our families very late, which did not contribute to the stability of our marriages, particularly mine. We were young, arrogant, and thought we were hot shit. After all, we were going to the Moon! The Singin' Wheel was our private retreat until one of our bosses, Arnie Aldrich, ruined it all when he brought his wife, Ellie.

It wasn't a place for wives.

Crazy times abounded at The Wheel; streaking was in vogue and some of us, not to be left behind, joined the fad. One day, John Llewellyn stripped off his clothes upstairs in the Club and dove out the window, rolled off the roof of the small front porch onto the parking lot, and proceeded to run the distance down Highway 3 toward the main intersection of Webster. It didn't take long for the Gestapo to roar to the scene. John dashed back to The Wheel's always-crowded parking lot, and hid between cars while he pulled on his clothes, which we had tossed down to him. As he sauntered back in through the front door of The Wheel, the cops turned into the parking lot in search of the reported streaker.

Only a few of the astronauts, most notably Jack Schmitt, Charlie Duke and Freddo Haise, would join us flight controllers in the Wheel for a "pre-mission briefing." Most of the astronauts were pretty oblivious to the operation of the Mission Control Center and the flight controllers and technicians who toiled to ensure their safety during the missions. To them, Mission Control was just part of NASA's giant infrastructure on the ground. At any one time during an Apollo mission, there were more than 400 people working in the building that housed the Control Center. Only Jack Schmitt made the effort to meet most of them. Before Charlie Duke flew on *Apollo 16*, we (flight controllers) imbibed some brews with him and shared dirty jokes. Charlie was a champion poor-taste joke teller and some of his tellings had us falling out of our chairs. As a matter of fact, during one of his lunar strolls, Charlie, observing the rugged Moonscape, commented to John Young, "That really looks bad." He confirmed later that it indeed was the punch line of one of his nasty jokes. To those of us listening in, it was a great insider moment from the lunar surface, all the more so because no one else had a clue.

Sadly, the Singin' Wheel closed after the completion of the Apollo Program and the death of its owner. It reopened as a short-lived German restaurant, after which it was permanently shuttered.

For us flight controllers, the Singin' Wheel was what Pancho Barnes' Happy Bottom Riding Club was to the Edwards AFB test pilots. Unfortunately, the expansion of the air base's boundaries and a fire put an end to the pilots' well-known watering hole in the early 1950s.

A place to have fun!

— Chapter 37 —
Flight Control: The New Mistress

Deanna and I had married young, but never grew together. It was easy to ignore my unhappiness with the marriage while busy full-time with school and my engineering career startup. While I was enjoying immersion in my exciting new job and environment, my marriage and my relationship with my children were going nowhere. In retrospect, I had traded the distraction of school and work for my new mistress, Flight Operations. Deanna suffered all this in silent acceptance. The marriage finally succumbed totally in 1972.

> *I never had the emotional maturity to be a husband and father, and my lack thereof was a major contributor to the marriage failure.*

Although it was sad to admit failure and get divorced, I was selfish because it meant that I could pursue the bachelor's lifestyle that I had missed out on as a result of marrying so young. I was mindful of my financial responsibilities to my children, but pretty much ignored their existence. Shelli grew up to be honest and self-reliant, and at age 12 was able to save her baby-sitting money to buy her first horse, which she kept for the next twenty years. She married Mike Kropp who brought her all the happiness for which she ever hoped. Scot developed into an honest man with a strong work ethic. He learned the roofing business through hard, backbreaking work, and he now owns a small roofing installation and repair business. Mark, the older twin by seven minutes, like Scot missed out on having his father present during his formative years. Coincidentally, he too went into the roofing business, and after years of hard work, became an accomplished salesman for a large roofing company.

Later in life, Shelli made a concerted effort to draw me close and was ultimately successful. For a time after that I tried to connect with the boys, but, because of my own failings, I was not successful.

> *I suppose males learn to be little islands early in life; I certainly had.*

Daughter Shelli Liebergot Kropp

Christmas, 1986: L-R: Son Mark, Sy, son Scot

Mission Control Center Building 30

— Chapter 38 —
Mission Control: Origin and Fate

The hub for ground-based Apollo mission operations activities was Building 30, the Mission Control Center in the NASA Manned Spacecraft Center. The drab appearance of this three-story grayish white, windowless monolith gave no hint of the critical operations being conducted within. The 1st floor was dedicated to the Real Time Computer Complex (RTCC) and various communications functions. The 2nd and 3rd floors each contained a Mission Operations Control Room (MOCR, pronounced MOKER) and Staff Support Rooms (SSR). Either MOCR could be used for a single mission, or in rare circumstances, they could be used simultaneously, such as was employed when the early unmanned Apollo missions overlapped the final few Gemini missions. During this period, the requirement for two groups of flight controllers to operate two completely different types of spacecraft placed a strain on even experienced personnel.

The old Mission Control has been replaced by a modern facility in an adjoining wing that is the new control hub of flight operations support for the Space Shuttle and International Space Station. In the new parlance, the MOCR has become the Flight Control Room (FCR, pronounced FICKER) and the SSRs have been replaced with the Multi-Purpose Support Rooms (MPSR). The old dedicated consoles have been superseded by state-of-the-art computer workstations that can be configured to any spacecraft system discipline with ready access to incredible processing power.

Mission Control that was exercised from the 2nd and 3rd floors during Gemini and Apollo left behind a rich history of accomplishments:

The 2nd floor housed MOCR-1. We had a flight control team in MOCR-1 for *Gemini 3* to check out the MCC. The MCC performed very well for *Gemini 3*, and the decision was made to control the rest of the Gemini missions from

Modern Shuttle Flight Control Room (FCR)

Houston. Saturn IB flights were all controlled from MOCR-1. There were four unmanned Saturn IB flights controlled from this floor prior to *Apollo 7*, the first manned flight. They were (in order): AS-201, AS-203, AS-202, AS-204L, this last being an unmanned LM flight that was later renamed *Apollo 5*. All four of the Skylab flights, namely the unmanned *Skylab 1* Orbital Workshop package, which was launched on a Saturn V, and the three manned flights of *Skylab 2*, *3*, and *4*, launched by Saturn IB, were controlled from this floor, as were the ASTP joint mission with the Soviets and the Skylab re-entry activities. This facility was later modified for the early Shuttle missions.

The 3rd floor (MOCR-2) – *Gemini 4* through *Gemini 12* were controlled from MOCR-2. All Saturn V lunar flights were controlled from the third floor. The two unmanned Saturn V flights AS-501 (later renamed *Apollo 4*) and AS-502 (later renamed *Apollo 6*) were followed in rapid succession by the ten manned lunar flights. The third floor was last used when the Shuttle carried secret DoD payloads.

The end of the golden era of manned space exploration "firsts" came with the closing down of the old Mission Control. The 3rd floor was decommissioned in late 1992, but because *Apollo 11* was controlled from MOCR-2 it has been reconfigured as it was for that mission and designated a national historic site.

When the 2nd floor was decommissioned in mid-1996, all the consoles were classified as "excess" equipment and stripped out. Sadly, no one knows what

Restored flight controller console in Aerospace Museum in Niagara Falls, NY

happened to most of them. I helped Mark Caterina obtain a Flight Activities Officer (FAO) SSR console. When mutual friend Fred Schoeller arrived at JSC to take possession of the 990-pound console, it was laying in a field, rusting. The property tech commented, "Who would want this piece of crap?" Mark restored the space relic and it was placed on display in the Aerospace Museum in Niagara Falls, NY, where he is an active, dedicated volunteer. This console has become one of the museum's most popular displays.

In the Apollo era, Building 30 housed approximately 146 consoles. Of these, only 40 have been tracked down – they are at:

  o 3rd Floor MOCR, Building 30, Johnson Space Center, Houston, Texas

- Aerospace Museum, Niagara Falls, New York
- U.S. Space & Rocket Center in Huntsville, Alabama
- Kansas Cosmosphere Space Center in Hutchinson, Kansas
- Smithsonian National Air & Space Museum in Washington, DC
- Tom Stafford Space Museum in Weatherford, Oklahoma
- Warhawk Museum in Nampa, Utah
- Michigan Space & Science Center in Jackson, Michigan
- Tulsa Air & Space Center, Tulsa, Oklahoma
- Omega LTD, Bienne, Switzerland
- JSC Public Affairs storage
- Grisham Middle School, Austin, Texas

and a few may remain to be found, but sadly the rest of these wonderful space artifacts have undoubtedly been reduced to scrap.

Each MOCR featured a solid walnut railing that stood out in stark relief to the twenty sage-green consoles. Jim Brandenburg, the Building 30 Facility Manager, rescued the railing from the scrap heap and had it sawed it into thin slices for mementos for those flight controllers who had worked there, such as myself.

MOCR-1 Walnut railing slice

I experienced a sense of loss when the Apollo-era Control Center was totally decommissioned. I'm certain that the Mercury Flight Controllers felt the same sense of loss when the Mercury Control Center at the Cape was shuttered, and Houston took over the Gemini missions.

— Chapter 39 —
A Little Tutoring Here: Mission Operations Control Room

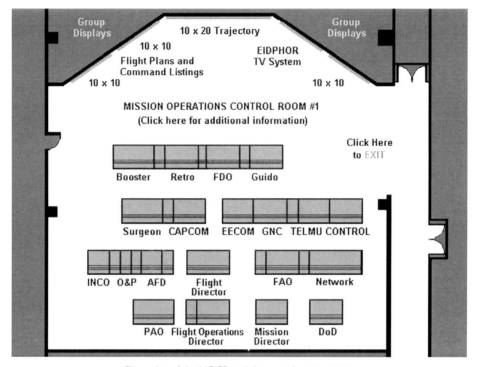

Floor plan of the MOCR and the console assignments

    The MOCR was laid out like a theater, complete to the five large rear projection screens, called the Group Display, at the front of the room. To the flight controllers, seated in four rising tiers, the wall-screens were incidental to the CRT (cathode ray tubes) screens mounted in the individual consoles along each tier, which presented information that the particular controller needed to make evaluations and make recommendations on matters concerning the flight crew and spacecraft.

    The flight control team consisted functionally of real-time command and control, systems operations, flight dynamics and planning, and was organized in a hierarchy of mission responsibility.

    Up to twenty-five flight controllers reported directly to the Flight Director who sat on the third tier towards the rear of the MOCR, and functioned as the "Head Flight Controller." The flight controllers in the MOCR were supported by flight controllers possessing greater individual expertise and analysis capability in their respective CSM Vehicle Systems, Life Support, LM Vehicle Systems, Aeromed, Flight Dynamics, and Flight Director SSRs. The authority hierarchy was such that the flight controllers in the MOCR were immediately responsive to the Flight Director and the SSR flight controllers were immediately responsive to their MOCR counterparts.

The front tier at floor level, was the lowest and the darkest, and was informally referred to as "The Trench." This was the location of the Booster and Trajectory guys. The second tier belonged to the Systems guys, Flight Surgeon and CapCom. Arguably, the tiers of consoles were arranged in hierarchy of mission responsibility. Although the Flight Director had absolute authority and carried total mission responsibility, he was supported in a staff advisory capacity by "upper management" who were located on the topmost row. *Where else?* Behind "management row" was a glassed-in seating area called the Viewing Room for up to 74 VIPs who, no matter their seniority, had only "observer" status.

Other operations support areas within the facility that were manned during mission or training activities included a Meteorology room, a Spacecraft Planning and Analysis (SPAN) room, a Recovery Operations Control room, a Simulation Control Area (SCA), and an Apollo Lunar Surface Experiments Package (ALSEP) room.

The MOCR was not a pleasant environment in which to work. The somewhat depressing lighting was low-level, indirect and shadowless, but it minimized reflection on the CRT screens. The screens themselves displayed vital mission data with low-resolution white characters on black, which invited eyestrain. In the late sixties and early seventies, smoking was a generally acceptable behavior, and the atmosphere was pungent with smoke. I was a habitual pipe smoker, and my console was messy with a large tobacco pouch, tamping tools, and a large amber ashtray that held the pipes and ashes. The headsets worn by the controllers were constantly a-buzz with essential communication. The first time I put on my headset, I was stunned and confused. It was a babble of voices emanating from eight to ten simultaneous intercom loops – how could I ever know what was important to my job and me? During busy periods there might be dozens of voices on the various loops. It was a recipe for headaches and tension.

The tension involved in working in the MOCR could be incredible. Although there were warn-

Large messy ashtray for my pipes

ing lights on the consoles that would alert flight controllers to developing problems, there wasn't enough room on any console for warning lights to cover every contingency. A controller had to watch the tiny figures marching across his CRT to cue him to a developing situation. Numbers, numbers, and more numbers – if one was wrong, it could indicate the onset of a severe problem. And the numbers were constantly updating, row-by-row. What might seem incongruous and raise concern one second could revert to normal the next when the data was updated. Perhaps the worrisome item of data had simply been corrupted in the complex communications chain from the spacecraft to the console.

Ratty data?

The flight controller listened to the voices in his ear while scanning the CRT. Much of what was being said in clipped, terse statements had nothing to do with his job, but the one piece of information that correlated with what his eyes were seeing had to be caught and analyzed. The flight controllers in the SSRs helped to sift the data, but it could still be overwhelming.

And when a controller identified a problem, he had to know when to approach the Flight Director about it. Too early, and it might be a false alarm. Flight wouldn't like that. Too late, and the opportunity for fixing or minimizing the problem might have passed – the problem might have grown to catastrophic proportions. Flight wouldn't like that either. Often, the difference could be measured in seconds. The advantage of listening to so many intercom loops was that each flight controller was constantly aware of the "mood" of the room. When one flight controller drew Flight's attention to the possibility of a problem in one area, the rest of us immediately started to consider how that would affect us, and so our alert level, always high, jumped even higher. It was not without its faults, but the system worked. It was a stressful environment, though.

> *The bottom line was that none of it mattered, because the MOCR was where it was 'happening,' and being there was worth any discomfort.*

— Chapter 40 —
Essential Equipment: The Flight Controller Console

The sage-green (the actual color) consoles at which the flight controllers worked in the MOCR and in the SSRs, and those in many of the support rooms, included one or more CRT screens and the necessary controls to display data on a number of different channels. The data could be the same as that displayed on the large group display screens in front of the MOCR, and other data could be "called up" by changing channels from the TV Guide or by requesting a specific data display. The displays and controls on these consoles and other group displays provided the capability to monitor and evaluate data concerning the mission and, based on these evaluations, to recommend or take appropriate action on matters concerning the flight crew and spacecraft.

The console of the Apollo CSM life support systems flight controller, called EECOM, had two screens on which real-time telemetry data was displayed. Although various data formats were available, the two displays used most frequently were entitled ECS Cryo and EPS High Density, which contained in excess of 200 individual parameters that were updated once per second.

Each console was equipped with an Intercom Keyset that provided the flight controller with the ability to communicate with virtually anyone in the Mission Control Center, other NASA centers, contractors facilities, or even the world-wide tracking network. Using a Plantronics headset, the keyset permitted talking on one intercom "loop" while simultaneously monitoring up to ten other loops, and simultaneous talk (conferencing) on more than one loop was also possible. Another significant feature was that the volume of the loop selected for TALK was automatically increased 3 db above the loops on MONITOR. This feature permitted the flight controller to select one loop, such as Flight Director, as a primary one while monitoring the others at a lower volume. A flight controller had to listen selectively for certain key words, particularly his call sign, such as EECOM. One of the physical consequences of having multiple intercom loops blaring in my left ear for many years is tinnitus, a constant ringing tone in that ear.

Small inconvenience for all the thrills.

The console CRT monitors were rack-mounted drawers that could be pulled out by handles on both sides for servicing. At times of stress, a flight controller would find his hands firmly attached to these handles, so they came to be known as "security" handles. Like the buzz on the loops, the number of flight controllers gripping their handles served as an indicator of how a mission was progressing. Gripping techniques varied: gripping a single handle could indicate the onset of a problem; gripping both handles was a sure indication of a developing, serious problem. The handles were in widespread use during the anxious moments of an emergency, as was the case during descent to the surface of the Moon on *Apollo 11*, and during the *Apollo 13* crisis *when I had a firm grip on both handles.*

As an aid in recognizing out-of-tolerance parameters and spacecraft events, groups of indicator lights were located along the "eyebrow" of the console. The lights on these EECOM panels were referred to as limit-sense lights. A limit-sense light illuminated whenever the parameter in question strayed outside of high and low limits, manually set for that particular parameter. One such panel had 72 lights which included 12 limit-sense lights for pressure, temperature, and quantity in each of the cryogenic oxygen and hydrogen tanks. In normal operation, the EECOM set fairly tight limits on the limit-sense lights in order to get an immediate indication of parameter variations. Consequently, it was not unusual for several limit-sense lights to be burning.

As was the case on Apollo 13.

In any of the great department stores of yesteryear you could find the "cash carrier" invention, which sent money in little tubes traveling by air compression to another location in the building where change could be made. The first mechanical carrier used for store service was patented by D. Brown on July 13, 1875, but it was not until 1882, when an inventor named Martin patented improvements in the system, that the invention became widespread.

The Mission Control Center was laced with a network of pneumatic tubing whose pressure system permitted the rapid transport of carriers (we called them "p-tubes" for the obvious reason) containing documentation to the consoles. "Comin' atcha" was the common announcement of transmission. The carrier's door was closed by a rather strong spring, whose force was forever scraping the backs of our hands. When first introduced, it used a more sensible spring-loaded-open door, and flight controllers were forever sending sandwiches and other non-paper items through the system. During one of the last Gemini flights Manfred "Dutch" von Ehrenfried sent a carrier full of pencils through the system and the carrier door opened en route, scattering dozens of "No. 2s" throughout the miles of tubing, bringing the system to a halt for hours. After that, the carriers were modified to operate in a spring-loaded-closed fashion.

The SPacecraft ANalysis room, SPAN, was established in Mission Control to provide a technical management interface between the flight control and the engineering worlds. It acted as a clearing room for all data requests or new procedures via a formal process using record forms called "chits." (Hungarian-born Mike Vucelic's accent doomed him to forever pronounce "chit" as "shit".) Requests were reviewed and expressed in proper technical language. Flight operations coordinators in the SPAN were represented by an astronaut, a flight dynamics flight controller, an EECOM and a GNC (Guidance, Navigation and Control) spacecraft systems flight controller.

Mission Control included two unique features designed with the comfort of the flight controllers in mind. The Flight Controller Lounge was a cipher-locked room in the operations wing of Building 30 that was a quiet secure place to have a meeting or to "decompress" before going home. Mini-bottles of medicinal brandy were dispensed by the flight surgeons, and were a special treat until Chris Kraft, the Director of Flight Operations, discovered an empty mini-bottle in a wastebasket and put a halt to the practice. The operations wing even had a dormitory across from the Lounge, which enabled us to be readily on call, if needed. For example, the lunar Ascent Team, of which I was a member as EECOM, was already in the building before the descent so as to be available whenever ascent from the Moon occurred. The Flight Controller Lounge was converted to a small cafeteria during Skylab, which started out with good intentions, but it was soon dubbed the "Roacheteria." Both of these flight controller facilities were finally torn out and converted to office space.

I guess the powers-that-be figured that Skylab flight controllers would not need such home-away-from-home comforts.

On each floor of the Control Center there was a block of 48 small cube-shaped lockers, indistinguishable except by a numbered metal label. Each console discipline was assigned one locker, which was secured with a combination lock. Flight controllers stored their headsets in these lockers to prevent loss or forgetfulness. However, if you're the Boss, ... Chris Kraft forgetfully left his headset on his console after a long day of simulations. When he came by to retrieve it the next day, it was gone. How did the Great One handle the loss? He merely placed a 3x5 card on a bulletin board in the hallway outside the MOCR. The card read simply: "Please return my headset — Chris Kraft." It was back on his console the next morning.

I'm sure someone counted their lucky stars.

— Chapter 41 —
The Flight Controller: Apollo Era

The Apollo flight controller was part of the continuing evolution of the tradition begun by Chris Kraft during the Mercury and Gemini programs. The typical flight controller was young (at 33, I was among the oldest during *Apollo 11*), male, assertive to aggressive, certainly arrogant, and willing to make decisions under pressure; this last attribute was mandatory. Unfortunately for some of us, the aggressive attitude that we adopted and wielded at work would last a lifetime.

I suppose that in comparison to our hippy contemporaries we looked a bit "nerdy," with our white shirts, thin ties, and pocket protectors. There was no long hair and pony tails for us; it was mostly "buzz cuts." And, of course, my thick-lensed glasses with dark black heavy frames reinforced the "engineer look."

We were encouraged to speak our minds. At one of the flight mission rules review meetings, INCO Ed Fendell was arguing with Flight Director Gene Kranz about the rules concerning the ARIA telemetry-relay aircraft for the entry phase of an Apollo mission. Ed wanted it set up one way and Kranz did not. The argument went on for a few minutes and finally an exasperated Fendell loudly told Kranz that "he had more console time than him!" hoping that would cause Gene to capitulate. But to no avail; Gene, who was our boss, simply said it was his way. Period. And Ed shut up.

Although the astronauts could fly their spacecraft and ultimately took all the risk, the flight controllers, with their greater in-depth knowledge of the trajectory and capability to monitor and analyze the operation of the spacecraft's systems in real-time, *controlled* the flight. Some of

1964: Sy Liebergot

us considered ourselves "ground astronauts."

> *Our absolute dedication to the mission, to the veritable exclusion of almost everything else, proved to be career-limiting and relationship-destroying for some of us.*

Ragging and joke-playing, often to excess, were commonplace occurrences. Large egos were unwelcome and "Loose Threads" were fair game for pulling. Hershel Perkins, a LM Systems flight controller, was the resident practical joker, and he was always on the alert for new opportunities. One of his victims was Dennis Webb. Dennis joined the Flight Control Division as a young co-op engineering student. He was foppish in dress and managed a professorial countenance and calm manner. But Dennis was naïve, and Hershel, sensing fresh meat, called him on the telephone at his desk, identified himself as a telephone technician, and informed him that the lines were dirty and in need of their periodic cleaning. He directed Dennis to place the receiver in a wastebasket so that the dust would be contained, and Dennis dutifully complied. What a hoot! As the years passed, Dennis ultimately became "a mover and a shaker" manager in the upper echelons of the Mission Operations Directorate. He still dresses and looks the same, just older. Dennis always reminded me of Bob Kelly, an eccentric physicist I knew in the Army, who owned a 1935 Ford that I nicknamed "Puffed Cheeks" because of the distinctive shape of its front fenders. He drove it all over the U.S., content to never exceed 45 mph while everyone else raced by.

Although some astronauts served in the MOCR at the Booster console during Gemini, and most took turns as CapCom during integrated training simulations and actual missions, few of them fully understood the role and full capability of the flight control team.

The astronauts were based at the Manned Spacecraft Center, and they saw the Mission Control Center as something that supported them, but didn't need their attention and personal involvement. They rarely acknowledged the absolute dedication of the flight controllers whose *raison d'être* was mission success and, by implication, their safety. We were simply "Houston," which was our call sign, or in the case of Dick Gordon, Command Module Pilot on *Apollo 12*, "Hey Ground!"

Jack Schmitt, the geologist, who we referred to as Dr. Rock, was the exception. He spent more time in the Control Center than any other astronaut. He seemed determined to learn everything he could about the people and workings of Building 30. He spent time in the RTCC, the voice and telemetry processing areas, SPAN, and the SSRs, as well as the MOCR. He visited with the technicians and flight controllers, staying long enough on two occasions to accidentally spill coffee into the electronics of back room consoles, shorting them out. Jack would sit with various flight controllers, both in the SSRs and the MOCR during simulations and actual mis-

Astronaut Jack Schmitt in space suit

sions, learning who we all were and how we did our jobs. When Jack finally flew on the *Apollo 17* mission, he was able to respond to calls from Mission Control by acknowledging the flight controller by discipline and name. Other flight controllers and I formed a personal attachment to Jack. Gene Cernan, in his book, *The Last Man On The Moon*, tellingly referred to him as "a creature of the Control Center." When I tell personal anecdotes about Jack, I always refer to him "as the only geologist to get his rocks off the Moon." He immediately knows that a person has been talking to me when they repeat that phrase to him; kinda like friendly code talk.

Another exception was the magnificent performance of the flight control team in the rescue of the *Apollo 13* astronauts. As a result of that shared, intense experience, some of us formed lifetime bonds of brotherhood with the *Apollo 13* crew.

The original three Flight Directors, Chris Kraft, Gene Kranz and John Hodge, with their concomitant flight controller team colors of red, white, and blue, invented the Flight Director duties, thereby setting the example for their subordinate flight controllers. They were very bright men of uncommon cool under pressure who had an uncanny ability to assimilate information, sort it out, and make a clear decision. The flight control philosophy and the high standards they developed during the early years prepared other Flight Director candidates for the Apollo program. Subsequent Flight Directors chose yet more colors, and when those were exhausted, metals and gem names were adopted. They are now working their way through astronomical constellations. The early Apollo colors were officially retired and each proclamation is displayed on a wall of the historic MOCR.

How did we do our job? When a spacecraft problem occurred while telemetry was being returned, it was identified by flight crew observations and flight con-

The Famous Spilled Coffee Episode by Mel Brooks

It was during the *Apollo 16* Mission – I was at my console in SPAN. Jack Schmitt (better known as "Doctor Rock") was talking to me over my console. He was standing in a narrow walkway behind the row of consoles, leaning over my console. He had a cup of coffee in his hand. Someone squeezed past him, bumped him, and the coffee spilled into my console through the ducted top (for air cooling); sparks flew; smoke billowed out; the CRT went dead. I killed power to the console. Jack decided to look inside and survey the damage. He lifted the top access panel from the back (it was hinged at the front) tilting it forward. I don't guess Jack realized that we had this rack mounted on top of the consoles, on which we stored our operational documents. Well, he tilted the panel and the documents crashed down upon the row of consoles, thereby knocking over about 3 more cups of coffee, and spilling over lots of log books, flight plans, cal comps, etc. Instinctively, the occupants of the consoles leaped back to avoid getting hot coffee spilled in their laps. As they did so, they crashed into people standing between the rows of consoles, who in turn bumped into the people seated at that row of consoles. This resulted in about 3 or 4 cups of coffee being spilled along the 2nd row of consoles.

In about 30 seconds, Dr. Rock had started a chain reaction which just about wiped out the SPAN room. We spent the next couple of hours cleaning up. During the *Apollo 17* mission, we (SPAN) had the CAPCOM tell Jack Schmitt that the guys in SPAN wanted him to know that it was very peaceful in the SPAN, no spilled coffee, and that we were glad he was on the Moon.

With fond memories,
Mel Brooks, FOD SPAN Rep.

[Note: Fortunately, Bob Legler had Mel document this episode for the Apollo 17 15th anniversary reunion. Sadly, Mel has since passed away]

troller real-time data observations. Time permitting, a much more in-depth analysis would be made, involving a review of telemetry data received from tape recorder playback, trend analysis of actual and predicted values, a review of collected data by systems specialists, correlation and comparison with previous mission data, and an analysis of recorded data from pre-launch testing.

If a spacecraft problem occurred within a flight controller's monitoring responsibility, he had to be prepared to clearly and succinctly report the problem to the Flight Director, identify the likely cause, recommend a course of action, and assess the impact on the mission. Flight must then wed that report to other possible failure impact reports from other controllers, and then make a decision on the future conduct of the mission.

The systems flight controllers – the EECOM and GNC for the CSM and TELMU and CONTROL for the LM, together with BOOSTER, INCO and all their supporting troops in the SSRs – were the engineers of the flight control team. The unique preparation that these people experienced placed them in a position to influence spacecraft design and testing, and to contribute to the development of the mission documentation, such as flight crew procedures, mission rules, and the flight plan. This all-around capability was convincingly demonstrated in the development of the emergency power-down procedures for *Apollo 13*'s CSM and the re-power-up procedures which were utilized so successfully for that mission's entry and landing.

In 1964, while still a North American Aviation employee, I was able to influence the design of the CSM. I was creating an electrical schematic of the Sequential Events Control System (SECS) for the Systems Handbook that the astronauts and the flight controllers would use to troubleshoot onboard problems. The SECS was an absolutely critical spacecraft system that controlled vehicle separation functions and the parachute deployment system. I identified five crew switches that initiated critical functions that if they failed to operate, would result in death of the crew. I saw that each of the functions was completely redundant, even to there being two separate switch modules behind the Main Display Console, but actuated by a single "bat" handle. If the switch failed mechanically, i.e., "failed to operate," the critical event could not occur. This was a violation of the redundancy requirement of the contract North American Aviation had with NASA.

I composed a letter for Chris Kraft's signature, identifying the four switches that were Criticality One, i.e., if they failed to operate the result would be catastrophic. All hell broke loose. One North American Aviation Displays and Controls manager wanted to know "who the sonofabitch was that caused this to happen." Well, I knew that as a North American Aviation employee my job was in jeopardy, so I kept my mouth shut. Such was the thin line a contractor walked, even when crew safety was at risk. In the end, the switches failed the qualification testing; the failure mode with eerie prediction was "failure to operate." The production of these switches was awarded to another manufacturer and all the identified switches were completely dou-

> UNITED STATES GOVERNMENT
>
> # *Memorandum*
>
> TO : PA/Manager, Apollo Spacecraft Program Office DATE: Nov. 25, 1964
>
> FROM : FA/Assistant Director for Flight Operations
>
> SUBJECT: Apollo/Sequential System Single Point Failures — *Command Module (C/M)*
>
> 1. This letter is to bring to your attention the existence of certain Main Display Console (MDC) switches that represent catastrophic or unnecessary single point failures in the Apollo Sequential System. The single point failures are manifested by the fact that both System A and System B logic for a particular function are terminated in a single MDC switch which initiates the function. The specific switches involved are as follows: *of the Mission Event Sequence Controller (MESC)*
>
>   a. C/M - S/M separate
>
>   b. Logic arm
>
>   c. Pyro arm
>
>   d. Abort Mode switch
>
> 2. A discussion of each switch follows:
>
>   a. C/M - S/M Separate Switch
>
>   This switch is used to separate the C/M from the S/M during a return from an earth orbital or lunar mission. During this phase of the mission the only method of separation is manual (pilot) using this single toggle switch. The switch is a single throw, double pole toggle switch. Mechanical failure of this switch would cause the loss of the particular separation functions since both System A and System B terminate in this single switch. Failure of this switch would be catastrophic since the S/C has retrofired and is now committed to reenter. In addition, there is no other method of separating C/M from the S/M (no back-up mode) during this mission phase. The reliability requirement, as specified, calls for a failure rate no greater than 0.25 percent per 1000 cycles. It is the opinion of this office that this failure rate is intolerable.
>
>   b. Logic and Pyro Arming Switches
>
>   These two switches are used to arm the sequential system logic and pyro busses. These switches have the same electro-mechanical and reliability characteristics as the C/M - S/M separate switch. Mechanical failure of either of these two switches would be catastrophic to the crew since this would cause the loss of all separation, RCS and earth landing functions.
>
>
>
> Buy U.S. Savings Bonds Regularly on the Payroll Savings Plan

Draft of the memorandum authored by Sy and signed by Chris Kraft identifying the five critical CSM switches that were to be doubled (page 1 of 2)

bled before we went to the Moon.

The flight dynamics flight controllers were a real departure from us systems guys. The Flight Dynamics Officer (FDO, which was pronounced FIDO), GUIDANCE (pronounced GUIDO), and the Retrofire Officer (RETRO), resided in The

c. Abort Mode Switch

The Abort Mode Switch has two positions: LES and Tower Jettison/SPS mode. The Tower Jettison/SPS mode position of this panel switch controls two functions simultaneously:

(1) Normal jettisoning of the L/E tower and

(2) Enables the SPS abort/sep circuitry.

Systems A and B of the MESC terminate at separate poles of this switch position; this switch has a single actuating lever. Failure of this position of the switch to perate would render useless both System A and B relays that control tower jettison motor and leg bolts fire. Since the leg bolts could not be blown, the only recourse would be a dangerous abort.

3. Recommendations

Mechanical failure of any one of the following switches would be catastrophic to the crew:

(1) C/M - S/M Sep.

(2) Logic and Pyro arm

Also, mechanical failure of the Abort Mode Switch in the Tower/Jettison/SPS mode position would cause an unnecessary and dangerous abort. It is therefore recommended that these single failure points be removed from the spacecraft, and that a significant improvement in the reliability of these switches be effected.

Christopher C. Kraft, Jr.

cc: D. Slayton (Astros)
FA/S. A. Sjoberg
FA/R. G. Rose
FS/H. E. Clements
FL/R. F. Thompson
FM/J. P. Mayer
PS/A. Dennett

FF:SALiebergot:pf

Draft of the memorandum authored by Sy and signed by Chris Kraft identifying the five critical CSM switches that were to be doubled (page 2 of 2)

Trench. In the spirit of "friendly" rivalry, the flight dynamics guys referred to us systems guys as plumbers and electricians while they (the exalted Flight Dynamics Group) wrestled with the elegant difficulty of powered flight, requiring (to hear them tell it) an intimate understanding of orbital mechanics involving Newtonian and Keplerian physics. The FIDO and RETRO pondered the meaning of the radar and telemetry tracking data on their screens and evaluated it with regard to a predicted flight path and the limits. They calculated the dynamic maneuvers required to keep the spacecraft on course and were supported by an SSR team of trajectory and computer analysis experts. GUIDO ensured that the spacecraft knew its position in space and periodically updated the on-board computer to that end.

1969: The Apollo "Trench" guys even had matchbooks and business cards

The flight dynamics flight controllers were never reticent to speak of their importance and accomplishments (they even had business cards and matchbooks), whereas the EECOM systems flight controllers seemed content with the knowledge that their systems were "a-cookin' and a-perkin'" just fine.

Charlie Dumis was an easy going, slow speaking East Texas boy whose manner belied the fact that he was an extremely bright engineer. I could understand how easy it would be for someone unfamiliar with Southern dialect to equate slow speech patterns with dull wittedness. The hometown of the "Doom," (his nickname) is Wamba, Texas; try to find that place on a map! I delighted in kidding him that the road only led into Wamba, not out.

Charlie was so slow to speak, that I recall Larry Sheaks, a high energy ECS engineer, telling me that when he would call Charlie on the telephone, Charlie took so long to say hello (collecting his thoughts) that Sheaks would hang up, thinking that no one was there. Then he'd call back and the process would repeat itself.

Apollo 9 L-R EECOMs Clint Burton and Charlie Dumis

It was natural that good-natured rivalry would develop between the various groups in Mission Control. Charlie was a fellow EECOM. Unfortunately, EECOMs rarely got a chance to fully explain what they had experienced in a training simulation because the FIDO would drone on with his long-winded debriefing report. One day, while waiting to debrief his part of a sim, Charlie penned the following bit of doggerel:

The Debriefing

Quiet! My Gut. Don't growl and roll
We both share a common goal.
A briefing that is smart and cute
With a FIDO who is mute.

Quiet! My Gut. Don't boil and seethe
Hungry though you may be.
Learn to satiate your needs
On the hot air that FIDO breeds.

Be still My Ass! Don't squirm and burn
Maybe someday we'll get our turn
To return to FIDO, at least, in part
Some of his oral farts.

Bladder! Hold your potential puddle
I'll grit my teeth and curse this muddle
Of a debriefing. It surely must be
That FIDO relieves himself orally.

Ears! I Hear! Can it be?
An end to this agony?
But no! FIDO's mouth moves again,
And re-invents the wheel again, again, again, again...

Oh Pain! Be thee gone?
Have I survived? No broken bone?
It's Over! End, my sorrow!
At least, until this time tomorrow.

While attending one of the many Apollo mission preparation reviews, in this case a Flight Mission Rules review, Charlie also jotted down the following, rather blunt private complaint. Some of it is prophetic with regard to *Apollo 13*:

Musings of a Bored EECOM

Nobody really gives a hair-assed fuck about EECOM Mission Rules.

Very few conditions create an Abort situation and most of those that do are difficult to define. They are similar to an orgasm – you know when it has occurred, but it is difficult to describe.

EECOM systems are totally alien to most other disciplines.

The Trench, the GNC, The Flight Director, the flight crew and CapCom (astronauts), the Booster Systems Engineer, even the FAO, all deal in flight path, its changes, and the reasons to change the flight path. These subjects receive the lion's share of any general discussion (e.g., Mission Rules reviews).

The EECOM Systems are like the water that is never missed until the well runs dry. Yet, the EECOMs are required to attend briefings for hours on end for the honor of having the assembled multitude listen to our offerings for a few precious minutes. Frequently, the long-winded crowd who so recently pissed away most of our day is excused from even listening to our few minutes.

Flight Directors generally don't understand EECOM systems (exception Kranz).

Chris Kraft in his Mercury hat

They start out with the intention of learning and understanding these systems, but once they find the systems defy category regimentation, they tire of listening to discussions of them.

The EECOM systems are like anchovies in that anyone can consume them, but far less people have a taste for them.

Christopher Columbus Kraft, Jr. was held in high esteem by us flight controllers. In fact, one of our nicknames for him was hay-sooce (Jesus), not to assign him deity status, but in recognition that he was the original Flight Director, who established the manner in which in manned space flight operations were conducted. He had an intuition for doing the right thing and though he always seemed to maintain a degree of aloofness, he was able to clearly communicate with those around him. He was considered one of the most powerful men in NASA; if he was behind you, you had as much leverage as you needed, if he was against you, you were dead meat.

Kraft carried a "big stick." When Chris was Director of Mission Operations, and later Johnson Space Center Director, he could make things happen; you just didn't want to anger him. If you argued with Chris on a subject with which he disagreed and it angered him, he would become impatient and might call you "young man." It would be in the context of that you were "dead," your career was over, you were a non-person. He had the power to do that. And I did that very thing on one occasion when, as AS-501 Assistant Flight Director (AFD), I over-zealously argued with him the need for certain pre-launch instrumentation for launch go-no go. Chris had little patience for the subject and cut off the discussion with, "Young man, I can't do anything about a cancer." *I had stepped over the line.* The Kraft hammer struck when my next promotion was reviewed by the Senior Promotion Board which was chaired by Chris, and he flatly rejected my application. Fortunately, my branch chief, Joe Roach, who was a member of the Board, pleaded my case and was able to convince Chris to overlook my youthful zeal and allow my promotion – I had been "returned from the dead."

Chris could make things happen. I observed him at a top management mission review around the giant mahogany table in the 9th floor conference room of Building 1. All the center senior directors were arguing about a subject, considering what action should be taken and how much money to allocate. The discussion went around the table many times until a consensus formed. At some point in the discussion, Chris had decided that he didn't like the direction in which it was going, but rather than debate, he remained silent until everyone else had made their case, then dropped his bomb with a simple statement, "I don't necessarily agree with that." With a few more insightful remarks, the whole discussion was re-launched and everyone would have to keep going until he was satisfied.

He had an uncanny knack for doing the right thing. He couldn't be bulldozed into making a poor decision. But he would wait until everyone else had their say, and then he would strike and tell them that they were wrong, and he had the sheer sense of presence to make them think again. It was magic to see him do that.

Odd that Chris always remembered my name.

Eugene F. Kranz was a genuine leader of people. His dedication to the job inspired us to put our all into being flight controllers and he kept us buoyed up. He truly

Gene Kranz in his classic vest

got out in front and led us Apollo era flight controllers to the Moon. He was an eminently approachable man. When he was on duty, we addressed him as Flight, and at other times he was either Gene or simply Kranz, and if we addressed him by his surname it was without rancor. During the Gemini program, he was affectionately nicknamed "General Savage" after the hard-driving general in the movie and TV series *12 O'clock High*; because he was never at a loss to inspire and set down written standards for his troops.

In order to formalize flight control philosophy, he compiled the Foundations of Mission Control:

The Foundations of Mission Control

To instill within ourselves these qualities essential for professional excellence:

Discipline ... Being able to follow as well as lead, knowing we must master ourselves before we can master our task.

Competence ... There being no substitute for total preparation and complete dedication, for space will not tolerate the careless or indifferent.

Confidence ... Believing in ourselves as well as others, knowing we must master fear and hesitation before we can succeed.

Responsibility ... Realizing that it cannot be shifted to others, for it belongs to each of us; we must answer for what we do, or fail to do.

Toughness ... Taking a stand when we must; to try again, and again, even if it means following a more difficult path.

Teamwork ... Respecting and utilizing the ability of others, realizing that we work toward a common goal, for success depends on the efforts of all.

To always be aware that suddenly and unexpectedly we may find ourselves in a role where our performance has ultimate consequences.

To recognize that the greatest error is not to have tried and failed, but that in trying, we did not give it our best effort.

* * *

Chris Kraft's approach to mission control was simple and concise:
- *Crew safety is the first priority.*
- *If you don't know what to do, do nothing (you may make it worse).*
- *Readily admit that you don't know something, when you don't.*

Gene seemed always to be seeking new ways to motivate. He issued an edict that the words "Tough & Competent" be written on each blackboard in the Flight Control Division offices. He would conduct whirlwind visits (in reality, they were nothing less than inspections) to all of our offices to maintain contact with everyone, to deliver well-chosen motivational remarks, and, I believed, to ensure that "Tough & Competent" was in full view on the blackboards. When he went on these wild tours, we characterized him as the "White Tornado," white being his team color. He had so much energy that each time he stopped smoking he would drive us nuts, especially when he started clicking his retractable ball-point pen during meetings. When the LM call sign was changed to TELMU, he never mastered its pronunciation, and always pronounced it TEL-eh-MU, which drove the TELMUs nuts.

Gene, ever the true believer in tradition and motivation, had the Mission Control Patch created in 1973 by artist Bob McCall – if the astronauts could have a mission patch then so could Mission Control. The patch was developed for the Mission Control Team to recognize their unique contribution to the Manned Spacecraft Program. The patch was updated for Shuttle.

I tell you, there were times that just standing near him would get me juiced up, such was his energy.

When he was Chief of the Flight Control Operations Branch within the Flight Control Division, Gene continued the Assistant Flight Director (AFD) position into the Apollo Program that had been created for the Project Mercury and continued through Gemini. The position description simply stated: "The AFD is responsible to the Flight Director for assistance in the detailed control of the mission, and assumes the duties of the Flight Director in his absence." Gene was the first AFD (to Kraft) and the only AFD to become an active Flight Director through to the end of Apollo. There was no training or certification for this new Apollo position. In reality, it was Assistant TO the Flight Director, and no Flight Director, including Kranz, treated it any other way. I served as AFD on AS-501, and lest I suffer the delusion that it was more that a gofer-type job, Flight Director Glynn Lunney put me straight in no uncertain terms. We AFDs urged Kranz either to abolish the position or to make it something more substantial. He ignored the recommendations, and the position went away with the end of the lunar program. John Temple had the dubious distinction of being the last of the AFDs.

Mission Control Emblem

I was Operations and Procedures Officer on AS-202, an unmanned flight, and I observed how odd it seemed that John Hodge, the Flight Director, seemed indifferent and permitted AFD Stu Present to pretty much run the show. Stu was also a NASA pilot and had had aspirations of becoming an astronaut, but was too old. Only two AFDs went on to become a Flight Director, and that was Chuck "Skinny" Lewis, who was my Flight Director during Skylab, and Harold "Dragon" Draughon, during Shuttle.

We all had nicknames.

I believe that, at heart, Gene Kranz was a frustrated systems flight controller; specifically, an EECOM. He loved systems and the systems guys. On one occasion, he told me, "I always believed you EECOMs had more fun." Kraft was at a different level, but Gene loved to talk to systems guys like me. In fact, he installed our back room loops on the Flight Director's console intercom keyset so that he could listen to our jabber about systems. He never used what he overheard against us, or for his own benefit; he just loved to listen.

Gene demonstrated organizational principles that I have carried forward in my life. If you're going to conduct a meeting, you must be the best prepared person in that room; you must overwhelm them with paperwork and data. That's how Gene conducted his professional life. He knew what he didn't know, and if he needed to know something that he didn't know, he studied until he knew as much as anyone with whom he had to discuss it. Where Kraft was absolutely brilliant and did it by inspiration, Kranz did it by 90 percent perspiration, working at it, loving the details. The further into the details you get, the more decisions you have to make, or the decisions are more technical, or more difficult to make because you've taken on so much data to consider, but Gene thrived on this. Chris Kraft and Glynn Lunney weren't like that; besides being very smart, they were more intuitive. Said Kraft of Kranz in his book *Flight*, "… it was sometimes scary to remember that he was human."

Having no assignment during the ASTP mission, he sat with me at my console as my "p-tube operator," so at his retirement party I presented him with a p-tube from my console, which I had decorated with red, white, and blue stripes. Gene stuck around through a good number of Shuttle flights before retiring in 1994. He has since written his memoirs, Failure *Is Not An Option*, and is now a sought-after lecturer.

Flight controllers studied hard, trained hard, and played hard. An essential part of our play was the post-mission celebrations. One of the flight controller traditions was mandatory attendance of the mission flight crew at the post-mission "Splashdown 'Debriefing' Party." The "Debriefings" were held in the biergarten (beer garden) of the Hofbraugarten, a German restaurant in Dickinson, ten miles south of MSC. It was owned by the Papach family who immigrated from Austria through Mexico to the U.S., and was managed by a family member we called Father George, an ordained minister (non-practicing). He loved having us patronize his restaurant, even allowing us to store our personal "Mission Mugs" on shelves in the biergarten bar. These were the beer mugs with which we would celebrate. Needless to say, after a while the celebrating would get pretty rowdy with a hundred or so young men drinking beer. Soon there were wrist wrestling and leg wrestling challenges issued and taken up. Kranz and astronaut Jack Schmitt could be found in the middle of the fray, vying to be the ultimate winner.

Can you say esprit de corps?

Sy's Flight Controller Mission Mug – back (left) and front

Mission mugs on shelves in the Hofbraugarten biergarten

— Chapter 42 —
Embracing a Tradition: Becoming an EECOM

EECOM is an acronym introduced for Project Mercury. It originally stood for Electrical, Environmental and COMmunication systems. This historic radio call sign is unique to manned space flight operations.

The EECOM position in the Mission Operations Control Room (MOCR) in the Control Center is steeped in tradition. Men who sat at the EECOM console during Apollo inherited standards of professional behavior and performance that gave newcomers to the position some pause. Ongoing Gemini flights and the addition of Apollo operations required training of additional EECOMs, subjecting them to a steep learning curve.

The Apollo EECOM Flight Controller was responsible for the life support systems of the Command and Service Module (CSM), which amounted to roughly half of the systems of that spacecraft. These were the systems that provided all electrical power and its distribution, heating and cooling, cabin atmosphere pressure control, breathing oxygen, cryogenic hydrogen and oxygen for fuel cell electrical power plants, the sequential system that controlled the separation events and the parachutes, and many of the mechanical systems. The EECOM was also responsible for the CSM communications system through *Apollo 10*, thereafter the responsibility was moved to a new console: INCO. The EECOM position responsibilities became further diluted on the Shuttle program.

The Flight Directors historically saw EECOM as a catch-all: if a function didn't fit anywhere, then it belonged to EECOM. An EECOM was willing to take responsibility for situations or anomalies not claimed by anyone else. "That's yours, isn't it EECOM?" was a common Flight Director query. Consequently, an EECOM was required to develop a broad understanding of the spacecraft systems and their operation.

EECOMs were expected to be honest and forthcoming, and on occasion they would get "sandbagged" by another flight controller. I must admit that some CSM GNCs were adept at this. Perhaps a definition of sandbagging is in order: *To delay the admission of a problem in one's area of responsibility until someone else admits to a problem first. Then and only then, chime in with your problem or problems, thereby not looking as bad as the first reporter.* At the Cape, this was a favorite tactic during launch countdowns when the CSM had problems, but reporting was delayed until hopefully the booster guys would report their problems first. It was a game of space chicken.

When I rolled into steamy Houston in 1964, the Mission Control Center was under construction, so the flight control team, gearing up for Gemini, used the old Mercury Control Center at the Kennedy Space Center. I traveled to see them training there, but sadly, I didn't really appreciate the historic value of my single visit. However, I was assigned to Apollo, and had my first taste of mission operations as a Sequential Systems Flight Controller for the first unmanned CSM test flight, Apollo Saturn AS-201. In 1966, I became a NASA employee and was quickly assigned as Operations & Procedures Officer (O&P) on the second unmanned CSM test flight, AS-202. I further got my flight controller feet wet as AFD on AS-501, the first Saturn V test flight.

Sy sitting at console in the Mercury Control Center at KSC

The training of the Apollo systems flight controllers was both extensive and intensive; it consisted of classroom training, studying schematics, design, documentation, discussions with the designers, hands-on training during simulations, and real mission experience.

After *Apollo 8* was assigned to go to the Moon, some of my training to become a flight controller was by "trauma," i.e., sink or swim. As an EECOM

Mercury Control Center building

trainee I was "side-saddling" with experienced EECOM Rod Loe during a full day of *Apollo 8* entry phase simulations. The first simulation and its debriefing were completed, and while the computers were being recycled for another run, Rod stood up, unplugged his headset from the console and casually remarked that he was "goin' for a cup of coffee and would be right back." I sat there, comfortable (a euphemism for fat, dumb, and happy) with the knowledge that all I had to do that day was to stay plugged in and just listen to the proceedings. However, time passed, and when the countdown to the commencement of the next run started, Rod was nowhere in sight. My heart pounding, I squirmed in my chair, my head swiveling around, looking for Rod, expecting to see him rushing back with his coffee, but he didn't appear so. I was faced with a choice – I could sit like a dummy and jeopardize the training run or I could jump into the simulation and do the best that I could. It was sink or swim time, so I moved over, plugged into the prime position, and ran the rest of the day. Of course, it was a ruse. The Flight Director and others who were in on Rod's scheme were quite amused at my discomfort, *but it had been a test, and I passed.*

A special relationship existed between the MOCR flight controllers and the members of their support teams in the SSRs located off to one side and across the hall from the MOCR, or the "back room" in the vernacular. Each MOCR flight controller had his staff of flight controllers, each of whom had had a more detailed knowledge of some specific systems in the spacecraft than he did. During Apollo, it was common for the Front Room flight controllers to be NASA employees and those in the back room to be mostly people employed by the Apollo contractors; however, that's the only distinction that was ever made. The training, dedication, and personal performance standards were the same for everyone.

> *They were brother flight controllers, united in a common goal: to place men on the Moon and bring them safely back to Earth.*

However, sometimes being a "back room guy" could try one's soul. Vern Howard was a young Air Force captain who had been assigned to mission control duty in Houston. Gaining operations experience was a prerequisite to his future assignment as a SAC (Strategic Air Command) ICBM launch controller in one of these holes in the ground somewhere in Kansas. He was a soft-spoken, polite and deeply religious person not given to profanity or even to raising his voice in anger.

During a day of *Apollo 7* simulations, I was working in SPAN and monitoring the systems flight controller intercom loops. Vern was the ECS flight controller in the SSR supporting Charlie Dumis, who was soloing as EECOM for the first time. The training guys created an anomaly in the cryogenic storage system to test the new EECOM. Both Vern and Charlie were well versed in the cryogenic storage system and Vern quickly reported the failure and followed up with a recommended course of action. Charlie, who had also seen the problem and was apparently pondering the situation, did not acknowledge him. Vern called, "EECOM, did you hear me?" After a pause he tried again "EECOM, ECS, did you copy?" Still no response from Charlie. Again Vern repeated his request – and again without eliciting a response. Each time Vern made his request, his voice became uncharacteristically louder, matching his growing frustration. We SPAN room guys monitoring the one-sided conversation, found Vern's dilemma entertaining and we wondered whether he would finally be provoked to curse. Each time he called Charlie and received no answer (there was nothing wrong with the communication system), we just knew he was at his wits' end and was ready to hurl some epithet at him, but he did not. We were on the edge of our seats with the expectation of a major behavioral breakthrough. Finally, with great exasperation in his voice, Vern practically shouted "EECOM, ECS!" Now it would happen, we thought, now the pure Vern Howard would be given to swearing! Again no response from Charlie, who apparently was still pondering. Vern sounded like he had reached the end of his patience and was ready to go over the edge – it was now or never – but all he could muster was "Good *grief*, EECOM!!" Damn, Vern was totally incapable of uttering a swear word! Disappointed, we resumed our monitoring of the sim. Vern finished his tour in Mission Control and did indeed get posted to silo duty.

The training guys were special people, but just as the astronauts received more attention than the flight controllers, the flight controllers received more attention than the training guys. In order to create and conduct the simulations to train and test the astronauts and us flight controllers, they had to know as much, indeed more than us. For a long time, we had an adversarial relationship with them because they were always out to "get" us. One training individual could be seen through the glass partition of the Simulation Control Area (SCA) on the right side of the MOCR clapping his hands in perverse glee when he was able to "zap" a flight controller. Oxymoronically, his nickname was *Good Will Murray*. Sometime later in Apollo, this adversarial relationship between the simulation guys and the flight controllers became cooperative; we became more of a team, and we were all better for it.

But they were still out to get us.

I recall an integrated (with astronauts) simulation training run for *Apollo 13*. While his buddies were tromping around on the lunar surface, the Command Module Pilot (CMP) was tooling around the Moon in a 60-mile circular, two-hour orbit, and as he passed behind the Moon, contact was lost with the Earth for 46 minutes. As usual, at the moment of Loss Of Signal (LOS), the data would get ratty. I'd have to decide which of the ratty data readings were for real, and which I could ignore, and as I watched I thought I'd seen the cabin pressure drop a blip then freeze completely. I called back to ECS in the SSR to ask if he had seen the cabin pressure blip; he offered that it had to be ratty data. I agreed, and suggested that we get some coffee while we waited for Acquisition Of Signal (AOS).

The CSM made contact with MCC right on time and I stared incredulously at the data screens. The cabin pressure was at zero – exposed to the full vacuum of space – and the CMP was sealed in his space suit, and asking for advice. I'd been had, BIG time. The sim guys had given me an irreparable cabin failure right at LOS hoping that I would miss it and although I'd caught a glimpse of it, I had dismissed it as ratty data. The sim guys had shown that we were seriously susceptible to such a failure, so we spent a considerable amount of time creating procedures for a "lifeboat" situation. How were we to transfer the CMP into the LM after the crew came back up from the surface? All three crew would have to live in the LM, now. They would need the food and water that was in the CSM, but transferring it would involve repeatedly depressurizing and repressurizing the LM cabin. Its oxygen supply would soon be exhausted. Could we repress the LM from the CSM cryo tanks? After listening to us work through the implications of a CSM cabin integrity failure while the two spacecraft were still separated, Flight Director Gene Kranz called a halt, saying "That's enough; we got enough out of that." We had established that we probably could have lifeboat procedures, at least for that case. But how likely was it that we would lose the CSM? I mean, really! Later, I reflected that Kranz had been kind to me and had allowed me to learn lessons from my laxity; I believe Kraft would have thrown me out of the MOCR.

Anne Accola's NASA MCC security badge worn when she was a flight controller during Apollo 17

And I just know the sim guys all had self-satisfied grins on their faces.

The Shuttle Program brought significant changes to JSC in the form of an influx of female engineers. That sea change produced female flight controllers and Flight Directors. By the final Apollo mission in 1975, there were only 12 Flight Directors. This number has grown to a total of 58 Flight Directors since Red Flight (Kraft). The Shuttle Program produced many female flight controllers and five female Flight Directors. Anne Accola had the distinctin of being the first NASA female flight controller when she was assigned to the EVA surface Navigation position in the SSR during *Apollo 17*. She also was the first and only female Simulation Supervisor (SIMSUP, pronounced sim-soup), and she comported herself admirably. The SIMSUP was the orchestra leader of that team of training specialists known to us as *Those Who Were Out To Get Us*. Each specialist had the task of learning as much or more than the flight controllers, so as to develop training "scripts" that were mission scenarios perforated with "faults" that would test not only the spacecraft systems knowledge of the flight controller, but more importantly, *our willingness to make decisions.*

Some of the SIMSUPs were reluctant to execute scripts that might be too difficult for certain Flight Directors and perhaps cause embarrassment, but not Anne, who could not be intimidated. During the debriefing of a particularly difficult and involved shuttle simulation, the flight control team did not perform well and the Flight Director challenged the SIMSUP, who was Anne Accola, as to the validity of the case thrown at them. An argument ensued, finally brought to a crashing climax when Anne announced, "I would rather fall on my sword before I will admit that I'm wrong!" Some of her colleagues affectionately nicknamed her "The Iron Maiden."

Ready communication between the front room and the back room was critical in dealing with contingencies. There was no better example than the interaction between this EECOM and my back room ECS flight controllers George Bliss and Larry Sheaks and EPS flight controller Dick Brown, in our struggle to deal with consequences of the *Apollo 13* oxygen tank explosion.

An Apollo EECOM tradition was that if an EECOM looked especially good in handling a problem during a sim, he would always give credit in the debriefing to his back room, and if the back room screwed up, the EECOM would say that *he* had screwed up and apologize to the Flight Director. It was with sadness that I witnessed ignorance of this tradition during the early Shuttle flights when I heard some FCR flight controllers blame their back room for a mistake.

Milt Heflin was the only traditional EECOM (albeit on Shuttle) who went on to become a Flight Director, and he later rose to head the Flight Director Office.

With regret, I discovered that other traditions were passing. As the Shuttle Program came on line, I discovered that I no longer had the zeal for the real-time job that I had sustained for more than a decade, and so I "retired" as an active flight controller. I took a job working for the Shuttle Program Manager, ex-astronaut Brewster Shaw, which required that I work in the SPAN as a back room guy. When I arrived there to assume my new Program Office position, Bob Legler, a flight controller dating back to Gemini, warned me that I would hear and see significant differences from the Apollo days. I was skeptical, but he was not wrong. I listened on the intercom loop of a Shuttle Flight Director as he called his team flight controllers for a status check with the statement, "Okay, all you Operators, give me an Amber [status light] when you're ready." Thunderstruck and sputtering, I asked Bob, "When did the flight controllers become *Operators?*" Is it because all they do is operate their consoles? Do they no longer control the mission? Is the Flight Director now the *Chief Operator?*" Bob chuckled at my indignation. Later, I had an opportunity to confront a couple of the new breed who still called themselves Flight Director, and who ought to have known better, but it was to no avail. The flight controller is now an "operator."

My era was over.

— Chapter 43 —
Apollo: The Early Missions

It is important to note that these are my personal recollections of the missions in which I participated in some capacity. There remain unmentioned missions in which I had no role.

AS-201, launched on February 26, 1966, was the first test flight of a Saturn IB launch vehicle. It was only a 37-minute downrange lob with a Command and Service Module, but it was a test of the sequential system that controlled separation, attitude control, and parachute deployment. It was also the first test of the Service Propulsion System (SPS), the engine that would place the astronauts into lunar orbit and bring them safely home.

It was my first exposure to flight operations, albeit as a contractor in a back room capacity as Sequential Systems. An interesting scenario occurred on launch day. After protracted holds due to problems with the booster, the Cape finally called Glynn Lunney, the Flight Director, to call a scrub. So Chris Kraft, who was in the MOCR, set off for Hobby Airport to catch a plane.

1966: Kurt Debus

While he was on the road to the airport, KSC Center Director Kurt Debus, called to Lunney and asked in his German accent: "Flight, can 've' unscrub?" The problem had been rectified. Lunney figured we could handle it, so the count was resumed and we launched a few minutes later. The short flight was completed successfully with the spacecraft splashdown in the Atlantic. As was his practice, Chris called in just before boarding the plane to update Lunney on his progress, and was astonished to be told that we had launched a few minutes after he departed the Control Center and that the mission was successfully completed. He was fit to be tied; he was *pissed*.

AS-202, launched August 25, 1966, was the second unmanned sub-orbital flight of the Saturn IB, lasting an hour and a half. For the first time the CSM included a Launch Escape System, fuel cells and cryogenic reactants, a guidance and navigation system, S-Band communications equipment, SPS engine complete with a propellant gauging system, and the environmental control system. Four perfect firings of the SPS engine, a rigorous reentry to test the CM heat shield, and splashdown in the Pacific chronicled another successful flight in a series of unmanned test flights.

AS-202: Sy at the MOCR Operations and Procedures Officer (O&P) console. AFD Stu Present in background

As a new NASA employee, I experienced my first duty in the MOCR as Operations and Procedures Officer (O&P) for this mission. It was exciting to feel an integral part of the preparation, countdown and launch of the mission, all hour and a half of flight time. I was on my way as a flight controller.

It was heady stuff, this Mission Control.

AS-204 (*Apollo 1*) was our first tragedy. The surprise was that it happened on the ground rather than in space. On Friday, January 27, 1967, less than a month away from their planned "shakedown cruise" of the CSM in low Earth orbit, Gus Grissom, Ed White, and Roger Chaffee were killed when a fire broke out in the Command Module during what was considered to be a routine test. The spacecraft was mounted on a Saturn IB on Launch Pad 34 just as it would be for the actual launch, but the rocket was not fueled. The plan was to go through an entire countdown sequence in a "Plugs-Out" test.

The astronauts entered the capsule to begin the test at 1 p.m. A number of minor problems cropped up that delayed the test considerably, and finally a failure in communications forced a "hold" in the count at 5:40 p.m. At 6:31, one of the astronauts (probably Chaffee) reported, "Fire, I smell fire." Two seconds later, White was heard to say, "Fire in the cockpit." The Apollo hatch could only open inward and was held closed by a number of latches, which had to be operated by ratchets. It took at least 90 seconds to get the hatch open under ideal conditions. The cabin had been filled with a pure oxygen atmosphere at higher than normal atmospheric pressure for the test, and there had been many hours for the oxygen to permeate all the material in the cabin. The fire spread rapidly, and the astronauts had no time to get the hatch open. White Room technicians tried to get to the hatch but were repeatedly driven back by the heat and smoke. By the time they succeeded in getting the hatch open, the astronauts had already perished due to toxic smoke inhalation and burns.

There were muttered complaints from various quarters about the shoddy quality of the spacecraft and the punishing "Go Fever" work pace. Chris Kraft summed it up bluntly, "We were too much in a goddamn hurry!"

It was the first of two oxygen fire occurrences in the Apollo spacecraft.

The Apollo Program was put on hold while an exhaustive investigation was made of the accident. It was concluded that the most likely cause was a spark from a short circuit in a bundle of wires that ran to the left and just at the foot of Grissom's couch. The oxygen environment allowed the fire to start and the large amount of flammable material in the cabin spread it quickly. In such an environment, even aluminum will burn. In effect, it was a "flash" fire. The rapid increase in pressure had ruptured the cabin wall. A number of changes were instituted in the program over the next year and a half, including a new hatch design which could be opened outward in just 2.5 seconds, removing much of the flammable material and replacing it with self-extinguishing components, using (for the first time in a NASA spacecraft) an oxygen-nitrogen mixture at launch, and the recording of all

Apollo I exterior fire damage

Apollo I interior fire damage

changes and overseeing all modifications to the spacecraft design more rigorously.

The changes made to the Apollo Command Module as a result of the tragedy resulted in a highly reliable craft which, with the exception of *Apollo 13*, helped make the complex and dangerous trip to the Moon almost routine. The eventual success of the Apollo Program is a tribute to Gus Grissom, Ed White, and Roger Chaffee, three fine astronauts whose tragic loss was not in vain.

The mission was originally designated *Apollo 204*, because it was to be the fourth flight of the Saturn IB, but it had been informally referred to by almost everyone as *Apollo 1*, so it was officially assigned the name *Apollo 1* in honor of the lost crew.

After the investigation, Spacecraft 012, the *Apollo 1* Command Module, was placed in a storage facility at NASA's Langley Research Center.

Interestingly, when the AS-204 Saturn IB was finally launched, it carried the first Lunar Module on an unmanned test flight (and hence was designated AS-204L). It was almost as if no one wished to place another CSM on that particular rocket.

> *My friend Larry Canin, who was the AFD in the MOCR, coordinating MCC's end of the KSC pad test, told me he never got their screaming out of his head. Never. He has since passed away and hopefully the screaming has, as well.*

The recent tragic loss of the Space Shuttle *Columbia* and her crew of seven astronauts may well diminish the commitment of the United States to a manned space program. Our successful manned ventures into outer space have not been without sacrifices. The *Apollo 1* pad fire resulted from an ignored flammability problem - corrective action was taken. *Apollo 13*, though the astronauts were rescued, ultimately could be blamed on ground processing failures - corrective action was taken. The explosion of *Challenger* occurred during freezing pad conditions. Though the failure was specifically attributed to a solid rocket booster joint O-ring flawed design, it was, in fact, the result of NASA succumbing to an unrealistic self-inflicted launch schedule - corrective action was taken. Not to be overlooked is the effect of continuing budgetary constraints levied on NASA that has resulted in the utilization of 25-year-old technology and no replacement vehicle in sight. The root cause of the loss of *Columbia* has yet to be determined.

The debris from the Challenger accident, the spacecraft was transferred to permanent storage in abandoned missile silos at the Cape Canaveral Air Force Station, Florida. Several years ago, thanks to Lieutenant Colonel Johnny Johnson (USAF, Retired), I had the privilege of visiting the site. It was a most sobering experience.

AS-501 (*Apollo 4*), launched November 9, 1967, was the first full-up Saturn V launch vehicle with a fully outfitted CSM with an ad hoc Mission Control Programmer (MCP) installed to execute the functions that would later be performed by

AS-501: Sy at Assistant Flight Director (AFD) console

the astronauts. The flight objective was to duplicate the lunar injection profile, and to test the heat shield. However, since we did not want to send the spacecraft on a loop around the Moon, the S-IVB stage placed it into an orbit with a 10,000 mile apogee, and then the SPS engine was fired to propel the Command Module back to Earth at lunar-return velocity. The total flight time for this thoroughly successful test flight was eight and a half hours. With this one flight, the lunar program took a giant step forward.

My life-long disappointment is that because I was on duty in Houston, I never witnessed a Saturn V launch. However, I did visit the launch pad and was allowed to climb to the top of that monster and rap my knuckles on the very top of the Launch Escape Tower.

I was moved up to Assistant Flight Director (AFD) for this mission, and as such my pre-mission duties involved attending lots of meetings in which I represented the Flight Director, Glynn Lunney. In particular, I had to ensure that mission documentation was completed on schedule. The resident Marshall Space Flight Center (MSFC) Booster Flight Control Office was headed by Chuck Casey, who had recently

AS-501 Assistant Flight Director Sy Liebergot

AS-202: Chuck Casey (foreground) at Booster console

taken over the office from Scott Hamner. As I attended meetings with Chuck and his subordinates, I couldn't help noticing changes in his behavior: lack of clothing changes, weight loss, late nights at the office, sometimes sleeping in his office and a desire to know the tiniest detail of every aspect of the booster design. During simulations, and even during the actual launch, Chuck exhibited a reporting quirk that drove Lunney crazy: During the critical 10-minute launch phase, Lunney would periodically poll the MOCR team on their status. We would simply say, "GO!" but Chuck would always report, "GO, Flight," adding with a slight hesitation, "at this time." On flight day, the Saturn V performed flawlessly, but Chuck's deterioration continued toward a nervous breakdown. He had apparently convinced himself that the success or failure of the first all-up Saturn V test flight rested squarely on his shoulders, and his alone. Subsequent to the flight, he was returned to MSFC and given plenty of time off.

The AFD was held in low esteem by the rest of the flight controllers and had no responsible real-time mission role. Fellow flight controllers delighted in pronouncing it "aphid," a small plant-sucking insect. Lunney, who had nothing but disdain for the AFD position, told me in no uncertain terms there was "no way in hell" that I would sub for him during the mission if he was unable to perform his duties as Flight Director, and he advised me that after the flight I should seek a more meaningful job. I duly requested transfer from the Flight Control Operations Branch (FCOB) to the Systems Branch, managed by Arnie Aldrich, where I ultimately and proudly became an EECOM.

Thank you, Glynn Lunney.

Apollo 7 marked the Apollo Program's recovery. In the year since the pad fire investigation report was issued, all of the recommended spacecraft changes to the Command and Service Module had been completed. Remarkably, despite more than 21 months delay, we were still confident that we would be able to fulfill JFK's goal of landing a man on the Moon before the end of the decade. The familiar "Go Fever" had returned, but we were in much better shape to make that effort now.

Apollo 7, launched October 11, 1968, was the first manned Apollo mission. It flew a 10-day Earth orbit mission, which tested all the CSM spacecraft systems. Unfortunately, for some reason, the crew of Wally Schirra, Donn Eisele, and Walt Cunningham exhibited great disdain for any input coming from the ground troops, especially from the flight control team. Wally particularly so, which surprised us because it was a test flight, and they were in fact, the test pilots. Wally constantly complained about this or that system, and challenged any requests to deviate from the flight plan. A crew adopts the mood of its commander, and so Walt dismissed a test of the backup cooling system as "a Mickey Mouse procedure."

I was working in the SPAN room for this flight. It was embarrassing to listen to astronaut tantrums and insubordination. Glynn Lunney, the lead Flight Director, became so enraged by the astronauts' continuing unprofessional conduct that he directed the RETRO to target their splashdown in a hurricane that was close to Hawaii. Of course, that didn't happen. I suspect the primary reason for Wally's bizarre behavior was the deep distrust for the engineering community that he had developed as a result of the Pad 34 fire; after all, it could have easily been him in Spacecraft 012 on that fateful day.

The mission did have light moments, and one occurred during one of the astronauts' daily TV shows. During the program, nicknamed the "Wally, Walt, and Donn Show," the astronauts held up signs such as "Hello From The Lovely Apollo Room, High Above Everything." Another impertinent one asked: "Deke Slayton, Are You A Turtle?" In accordance with barroom tradition, adopted by the astronauts, he was obliged to answer, "You bet your sweet ass I am," or buy everyone within earshot a drink. "I have recorded my answer," Slayton responded after momentarily cutting off his microphone.

In 1962, Wally Schirra's Sigma Seven flight, introduced me to The Turtle Club when the following exchange was recorded 3 minutes into launch:
Deke Slayton: *Are you a Turtle, today?*
Schirra: *Going to VOX Record Only. You bet ... (correct answer was recorded)*
Slayton: *Just trying to catch you on that one.*

What I heard reference to was the International Association of Turtles that has its origin in World War II. History has it that some American pilots were relaxing in an English pub during World War II when they somehow got onto the subject of tur-

tles. They described a person who was clean-minded, usually minded their own business, but when the need arose, would stick their neck out for themselves and other people in need who had the attributes of a turtle. Sounds crazy unless you've been around drinking pilots. These pilots became the first members of the International Association of Turtles. Knowing the way of the military that they would be split up soon, and new members would be recruited, they came up with a way for Turtles to identify each other. When anyone asked a Turtle, "Are you a Turtle?" the proper response was "You bet your sweet ass I am." (YBYSAIA) Failure to do so required the delinquent respondent to buy everyone within earshot a drink.

While still at North American Aviation, I met George G. Gremillion, the local Imperial Turtle, who inducted me into his Turtle chapter. A while later, I was honored to become a member of The Turtles Outershell Division, headed by High Potentate Wally Schirra.

Throughout the *Apollo 7* mission the crew continued their rebellious behavior, but despite it the mission objectives were accomplished and the flight controllers and engineers added to their knowledge base of the Apollo spacecraft. However, as the end of the mission approached, the contumacious behavior was to continue. Wally had a head cold and didn't want to wear a helmet during entry. He feared damaging his hearing if he was unable to hold his nose and clear his ears in response to the pressure changes during the descent. The Flight Mission Rules called for wearing helmets during entry, and Kraft issued a direct order to Schirra, "Wear the helmet." Wally's response was simple: "Go away." Chris was angered by the lack of discipline because there was noth-

1961: The Turtle Club membership card - backside with questions

1961: The Turtle Club membership card - front

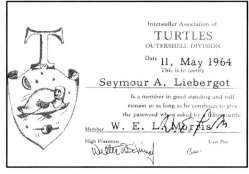

1961: The Turtle Club Outer Shell Division membership card

ing he believed in more strongly than discipline in the astronaut corps. We were all mightily relieved when splashdown in the Pacific brought an end to ten days of bickering. Wally hailed the mission as "101 percent success." It was essentially the same mission that was assigned to *Apollo 1*. Although Gus Grissom and Wally were contemporaries, there is no way that Gus would have given us such trouble.

Wally had let it be known that he would retire after this mission, so that may have been a reason why he felt free to let loose his frustrations. However, Cunningham's flight status was a different story. I watched an interviewer in the 1995 TV documentary *Moon Shot* query Chris about this and he replied, "What I said was that 'the only way Cunningham will ever fly again was over my dead body!'" After hearing that Kraft had decided that he wouldn't fly again, Walter Cunningham disputed that Chris had said that. Later, Walt called Chris to confirm if that was the case. Chris repeated his statement to him followed by, "You got it straight from the horse's mouth."

I'd been told that Wally Schirra had quite the overblown ego and one day I watched him and Al Shepard on TV try to "out knowledge" each other in commentating on one of the early Shuttle launches. At one point, Shepard called Wally what sounded like "Sky Ray" and my ears perked up. Sky Ray? Later, I had an opportunity to ask one of my ex-astronaut friends, Al Bean, about this, and he told me that during a press conference by "The Original Seven," a reporter, not knowing how to pronounce the name Schirra, addressed Wally as "Sky Ray," much to the delight of his fellow astronauts. Wally was not amused. From that day forward, he was privately referred as Sky Ray within the astronaut corps.

Sy receiving an award from Gene Kranz in 1968

— Chapter 44 —
Apollo Continued: The Lunar Missions

Apollo 8, the Christmas Mission, was launched on December 21, 1968. The 7-day mission manned by Frank Borman, Jim Lovell, and Bill Anders was a bold step toward the goal of a lunar landing. It was the first time men rode atop the huge three-stage Saturn V vehicle, and the first flight to take men not only to the vicinity of the Moon, but into orbit around it. This mission came about because the development of the Lunar Module was running behind schedule, and rather than delay the program, NASA, encouraged by the success of the CSM's shakedown test, decided to send a CSM out to the Moon to assess deep space navigation, tracking and communications.

It was a historic time for our country, and the significance was not lost on me, for we were ever mindful of the Russians' effort to be first to land men on the Moon. It was my first time at the EECOM console, albeit as a trainee. When *Apollo 8* disappeared behind the lunar disk, it was "pucker time" for all of us. For the spacecraft to achieve lunar orbit, the SPS engine would have to fire. We would know if events had gone according to plan by the exact time that the CSM came out from behind the Moon. It was right on time! We were in lunar orbit! I was so excited that without thinking, I stood up and shouted loudly, *"The Russians suck!!"* As time passed, I completely forgot my outburst, but Jack Schmitt had been present (where else would he have been than in the MOCR!) and he reminded me of it at a social gathering years later.

But I was correct. That Christmas we beat the Russians by a whisker because, as the recently released archives confirmed, they'd been working on a loop-around-the-Moon that would have allowed them to steal our thunder, but they had a run of bad luck and their mission, which would have launched a few days before ours, was canceled at the last minute.

Apollo 9 was launched March 3, 1969, as a 10-day Earth orbit mission by Jim McDivitt, Dave Scott and Rusty Schweickart, in order to qualify the Lunar Module for lunar orbital operations.

This was my first space flight to solo as an EECOM and I admit to butterflies as Charlie Dumis, who I was relieving, briefed me and departed. Charlie already had two missions of experience. Early on, the automatic pressure control system for the fuel cell hydrogen tanks failed. This necessitated manual operation of the tank heaters. We were only just learning how to deal with the characteristics of minus 420 degree Fahrenheit cryogenic hydrogen, and this was the first time that tank cryogenic pressure management had been exercised. Dumis' team had bumped the pressure high enough that we thought it would drop slowly enough to stay above the low-pressure alarm limit during the entire crew sleep period. I inherited the issue. As we said goodnight to the crew, Flight Director Cliff Charlesworth asked me how well we could predict the pressure decay characteristics of the hydrogen tanks. To this day, I don't know what prompted me to reply, "With incredible accuracy, Flight." Perhaps it had something to do with Cliff's laid-back personality that went along with his nickname of "Mississippi Gambler." One hour later, we had to awaken the crew due to an impending low-pressure alarm. Charlesworth was not happy with me.

EECOMs were resourceful fellows. About mid-mission, I discovered that I could create a "private" internal intercom loop out of the pre-launch communication loops to KSC that we had used for launch and were now out of use, and so I used this to ask my back room SSR guys "dumb shit questions" that I didn't want anyone else to hear. I named it "Bluebird." The first time I told Jim Kelly, my SSR EPS guy, to "Go to Bluebird," it sent Rod Loe, both our Branch Chief and a systems flight controller in SPAN, into a tizzy because he couldn't listen in. He hustled into the MOCR and demanded to know what this "Bluebird" was. Unfortunately, once my ingenuity was recognized, Bluebird became a standard loop, and so lost its original usefulness. "Bluebird" under different names, has survived to this day among flight controllers.

Apollo 10 was launched May 18, 1969. This 8-day lunar orbit mission by CDR Tom Stafford, LMP Gene Cernan, and CMP John Young qualified the combined spacecraft (CSM and LM) with an undocking, descent to the 50,000-foot point at which the powered descent would be initiated as if they were to attempt to land, and then a simulated abort, rendezvous, and re-docking. It must have been frustrating for Stafford and Cernan to fly so close to the Moon without actually landing, but this was only a dress rehearsal.

Fuel Cell 1 failed during Trans-Earth Coast as a result of a dead short circuit in the hydrogen separator pump motor. Dick Brown, my EPS back room flight controller, alerted me to the indications on our data screens. It was the first in-flight fuel cell failure. I reported the anomaly to Flight Director Pete Frank and gave instructions to have the crew check the appropriate circuit breaker. Cernan reported back that the circuit breaker had indeed popped out (open). I advised that it was to be pressed back in, whereupon Gene reported that when he did so, "it followed my finger right back out." Needless to say, I advised him not to touch it again and briefed everyone on the Flight Director's loop that the fuel cell was lost and that there was no mission impact, no power down required, nothing; ops normal. It was fortunate that this had not occurred on the way to the Moon, as the mission rules mandated that the loss of a fuel cell would prohibit undocking.

Immediately after I finished my explanation, RETRO John Llewellyn erupted from The Trench, leaned over my console and demanded, "What's this about a fuel cell?" I replied with exasperation in my voice, "John, I just finished briefing everyone on the Flight Director loop about the fuel cell problem." Whereupon he got in my face and told me in extremely crude terms that if I didn't personally brief him, he would kill me. I believed him; I did. John, though very bright, was a rough-edged individual given to plain speech. His hair trigger temper on occasion made him a heap of trouble. He was probably born 500 years too late; he would have loved to have been a Viking, storming ashore and plundering. In fact he saw action in the Korean War as a sniper and was a survivor of the Battle of the Chosin Reservoir (the Frozen Chosin). John was a damn good flight controller, though.

John Llewellyn holding forth at the Singin' Wheel

I was surprised to discover that Gene Cernan did not recall that first in-flight fuel cell failure on *Apollo 10* and did not note it in his book, *Last Man On the Moon*.

Apollo 10 revealed a flaw in the communications procedures. The EECOM had the CSM communications responsibility and the TELCOM (the LM's EECOM) had responsibility for the LM's communications. In lunar orbit, Stafford and Cernan entered the LM for systems activation and checkout. The mission rules required that the LM crew and the astronaut remaining in the CSM have two-way communication before the LM undocked and separated from the CSM. Things went to hell; neither spacecraft was able to communicate with the other, prompting Tom Stafford to coarse criticism. After a delay of more than an hour, two-way communications were restored. The problem was traced to an erroneous switch configuration called out in the spacecraft checklists. Quick correction of the problem was delayed because two different console positions were involved, which resulted in lengthy discussions and comparisons of the two spacecraft checklists. After the mission, Gene Kranz decided to remove communications responsibility from EECOM and TELCOM. He also decided to remove communications responsibility from the EXPERIMENTS console where control of the Lunar Surface and EVA communications equipment resided. Kranz consolidated those three responsibilities, including the instrumentation equipment, at a new MOCR console position named INCO (INstrumentation and COmmunications). The transfer of responsibility began first with EECOM on *Apollo 11* and was completed with TELCOM on *Apollo 12*. After *Apollo 12*, the TELCOM call sign was changed to TELMU.

I am eternally grateful to Ed Fendell, who took on the challenge and leadership of the INCO job and performed it so well. His operation of the Lunar Rover's remote camera on the Moon earned him the admiration of the German TV media, who presented him an Emmy-like trophy with his new nickname, Captain Video.

INCO flight controller Ed Fendell

Apollo 11 MOCR Identity Badge

Apollo 11 was launched on July 16, 1969, and by landing two men on the lunar surface and returning them safely to Earth, more than accomplished John F. Kennedy's objective for the Apollo Program. The landing by the Lunar Module *Eagle* on July 20 marked the culmination of all our efforts over the decade. While CDR Neil Armstrong and LMP Buzz Aldrin walked on the surface, CMP Mike Collins remained on orbit in *Columbia* to recover them on their return.

As EECOM on Lunney's Ascent Team, I had to be in the MOCR for the landing, so as to relieve my counterpart on Kranz's Descent Team, should an emergency liftoff from the lunar surface become necessary. I therefore had the privilege of being present as two astronauts in a spacecraft 240,000 miles distant prepared for the first landing on another planet. It all seemed a bit unreal.

As a specialist for the CSM, the EECOM had little to do during the descent phase of the mission; the main players were the Lunar Module Systems Flight Controllers and The Trench, namely FIDO Jay Greene, RETRO Chuck Deiterich, GUIDO Steve Bales, LM CONTROL Bob Carlton, and LM TELCOM Don Puddy. As a member of the Ascent Team, I was a bystander as the drama was played out. It was impossible to stay calm as Bob Carlton called out the rapidly decreasing LM fuel supply, and the momentary confusion caused by the LM computer alarms that put the onus squarely on GUIDO Steve Bales. Fortunately, Steve had Jack Garman, one of the most able computer experts in his back room cueing him, enabling Bales to shout his dramatic "Go!" and thereby become an *Apollo 11* legend.

> *At times, our back room gurus would make us MOCR guys look awfully good.*

I'm often asked what was the most exciting of all the lunar missions, and the response is met with surprise when I say that *Apollo 8* was the most exciting – the interviewer is invariably expecting the answer to be *Apollo 11*. Many other flight controllers are of the same opinion, because *Apollo 8* was the *first* time we sent men to the Moon. However exciting it was, *Apollo 11* was the expected result of the systematic progression of test missions.

Sy with his trademark black eyeglass frames. Picture taken during Apollo 11 Splashdown Party

When asked, "How'd we do it?" the response of RETRO Chuck Deiterich was succinct: "It was done by a bunch of smart guys that could think straight."

Apollo 12 was launched on November 14, 1969, carrying CDR Pete Conrad, LMP Al Bean, and CMP Dick Gordon. This second lunar landing mission was to demonstrate the capability to make a pin-point landing. The selected target was beside the crater in which the *Surveyor III* robot spacecraft had landed three years earlier.

As the pre-launch EECOM, I monitored all the systems activation and documented any off-nominal operations. The Flight Control Team now had a dedicated Launch Phase Team, which had specially trained to respond in the critical, time-compressed launch phase. It was to prove its worth during this mission's launch. After handing over to John Aaron, I sat aside in the expectation of another textbook Saturn V performance.

Chris Kraft had said that launch was always an uneasy time for him, and that he always looked forward to successful separation from the booster. On this occasion, his apprehension was further increased by the bad weather over the Cape.

> *Even though they were seasoned space veterans, Walt Kapryan and Gerry Griffin shared "rookie" status on this mission; Walt conducted his first launch as the KSC Launch Director and Gerry experienced his first as a Launch Flight Director. Post flight, they received a lot of well deserved kidding about the "rookies who launched into a thunderstorm."*

Lightning strike soon after Apollo 12 was launched

Apollo 12: Sy Liebergot enjoying post-landing celebration cigar in the MOCR

It was raining heavily when *Apollo 12* launched into an overcast of stratocumulus clouds with a ceiling of only 2,100 feet above the ground. Rising from Pad 39A at 11:22 a.m. EST in defiance of a mission rule, which said no vehicle shall be launched in a thunderstorm, the 360-foot-long Saturn V promptly vanished into the murk.

Observers then saw two bright blue streaks … right where the rocket had been 36 seconds after liftoff. It was hit by lightning at an altitude of 6,100 feet, but flew on unscathed. At 52 seconds it was hit again, and this time a reverse electrical surge took the CSM guidance system down and knocked all three fuel cell electrical power plants off-line. Pete Conrad reported, "Hey Gang, I think we've been hit by lightning!" He then read down a lengthy list of caution and warning lights that had illuminated in the spacecraft, no doubt thinking that he would have to abort. Looking at the scrambled data on his console TV monitors,

John Aaron recognized a pattern that he had seen during a pad test the previous year. He calmly issued directions to have the crew select an alternate telemetry instrumentation power supply in the Signal Conditioning Equipment: "SCE to AUX." When instructed by CapCom Gerry Carr, Al Bean responded, "SBE to what?!" After a repeat of the instruction, this time spelling out the acronym more slowly, Bean found the switch and moved it to the AUX position. With the data now making sense, Aaron was able to tell Griffin to have the crew reconnect the fuel cells, "one at a time." The tension in the MOCR was so thick that you could have cut it with a knife. Later, Gerry admitted that if John had told him to abort, he would not have hesitated to issue the command to the astronauts. John, ever humble, received a NASA medal for his part in saving the mission and coolness under great pressure. Once in orbit, the guidance system was recovered and *Apollo 12* flew a perfect mission.

I was in awe of John's performance, but was told later by Rod Loe that I was not the first person to have an inferiority complex over his superior abilities, "so get over it."

Apollo 13 MOCR about five minutes before the oxygen tank explosion. The top of Sy's head can be seen in line with and in front of Flight Director Kranz

— Chapter 45 —
Apollo 13: The Longest Hour

All flight controllers know that someday they will be in a position where their performance has ultimate consequences. — Gene Kranz

Apollo 13 was launched on April 11, 1970. Its Saturn V booster performed perfectly except for a center-engine shutdown during the second stage S-II burn, but the booster Instrument Unit compensated for this by firing the other engines for a longer time and so it had little effect. The mission was to be the third lunar landing, this time in the hilly region of Fra Mauro. The commander was Jim Lovell, who had already flown around the Moon on *Apollo 8*, and would be the first astronaut to return to its vicinity, this time with the intention of landing. Fred Haise was the LMP. The CMP, Jack Swigert, had replaced Ken Mattingly only a few days before launch, because Ken had been exposed to German measles. At the time, Ken strenuously objected to the decision, but is more philosophical about it today. After all, *Apollo 16*, to which he was reassigned, was successful.

At 55 hours into the mission, the three astronauts were about 200,000 miles from Mother Earth. The spacecraft had passed the Neutral Gravity Point, the distance where the gravitational pulls of the Moon and the Earth were equal, and so it was now under the Moon's gravitational influence and accelerating as it closed in.

The first seven hours of the shift had rolled by uneventfully. CapCom astronaut Jack Lousma radioed the crew, "The spacecraft is in real good shape as far as we are concerned. We're bored to tears down here." It would be the last time anyone would mention boredom for a long time! The flight crew was finishing up a TV broadcast in which they showed how comfortably they were able to live and work in weightlessness, but none of the news networks carried the TV feed live. Incredibly, after only two lunar landings, the news media had apparently decided that traveling 240,000 miles away from the Earth and landing men on an alien planet was old hat, and so

Apollo 13 MOCR Identity Badge

now relegated it to a minor story. They seemed to be determined to exercise their "expertise" in deciding what the public wished to read about or see. Lovell who had not been told that his audience was limited to the MOCR, signed off with, "This is the crew of *Apollo 13* wishing everybody there a nice evening, and we're just about ready to close out our inspection of *Aquarius* and get back for a pleasant evening in *Odyssey*. Good night." *Aquarius* was the LM and *Odyssey* was the CSM.

By this point in the Apollo Program, having had to respond to my share of in-flight anomalies, I was a pretty seasoned EECOM. The first in-flight fuel cell failure, the various heater and leak problems, and all the other little glitches with which I had had to deal, had all been within our preplanned contingency procedures and mission rules. They had occurred in a manner that allowed me to "work the problem" methodically, without significantly impacting the mission.

I thought I was prepared for anything – I was wrong.

It began during the final hour of my eight-hour shift; my relief EECOM, Clint Burton, was due to show up soon to begin our handover briefing.

I thought, "Everything is under control."

I had a couple of simple spacecraft systems "housekeeping" items for the crew to perform before they went to sleep. I told Flight Director Gene Kranz that I wanted the crew to terminate a periodic Entry Battery B charge and to "stir" the four cryogenics tanks. The rechargeable entry battery was one of three small batteries that would be the CM's sole source of electrical power during re-entry into the Earth's atmosphere. The two liquid oxygen tanks provided breathing oxygen and, together with the two liquid hydrogen tanks, supplied the fuel cells that produced electricity and drinking water as a by-product. The cryogenics were more like a very dense fog than liquid and in the weightlessness of space, the fog would stratify into layers of different densities, which caused the quantity gauging system's capacitive probe to give false readings. So prior to taking a quantity reading, two small fans were switched on in each cryo-

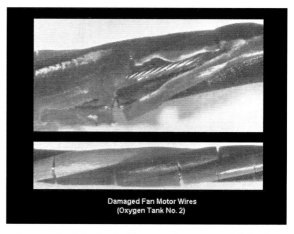

Damaged Fan Motor Wires (Oxygen Tank No. 2)

An example of the Apollo 13 cryogenic oxygen tank Teflon fan wiring cracked and bared after being subjected to excessive heater operation during the pad test

genic storage tank to create a stir, homogenizing the cryo. The tank stir was normally performed once a day, post sleep, but I requested an *extra* stir *before* sleep because the Oxygen Tank 2's quantity instrumentation had failed off-scale at the high end earlier and I wanted a more frequent quantity reading of Oxygen Tank 1. It was critical that we tracked the spacecraft consumables usage very closely. I could infer Tank 2's quantity from Tank 1, since they were plumbed together and fed simultaneously; a useful aspect of the design.

When Command Module Pilot Jack Swigert flipped the four switches to turn on the tank fans, no one knew that the wiring inside Oxygen Tank 2 was charred, cracked and bared. A spark in that wiring ignited a fire that caused the tank to explode within seconds. The force of the explosion was estimated to be a pressure of 60,000 psia and equivalent to 7 lbs. of TNT – enough to level a 3,000 sq. ft house. If you weren't looking at the exact data on the TV screen in those few seconds, you missed it. I did; we all did.

Suddenly, at 55 hours 53 minutes into the mission, the radio link crackled, as it did when communications were about to be lost; the explosion had blown off one of the Service Module's side panels, which had knocked the High Gain Antenna off angle as it sailed clear, forcing the communications onto the lower-power Omni antennas. Swigert reported, "Okay Houston, we've had a problem here." That was followed with reports of electrical anomalies. Flight Director Kranz asked me what I was seeing. Most of the data on my screens looked unreliable or was static, as if we had an instrumentation problem, and so I responded, "We've got some instrumentation, Flight; let me add them up." Within a minute, however, I began to hear other reports of problems on the Flight Director loop: INCO had a problem with the High-Gain Antenna; GUIDO reported that the computer had a "hardware restart"; GNC saw wild reaction control jets firings, jet valves closed; and onboard the crew was calling down indications of multiple failures with the Caution and Warning System alarm constantly sounding. It was unbelievable. Kranz queried me again and I repeated my initial impression, "We may have had an instrumentation problem, Flight." Talk about an understatement! Those words have haunted me since that time.

I had been sitting next to EECOM John Aaron when *Apollo 12* was struck by lightning soon after launching into a thunderstorm. Then, most of the telemetry data became static and it had been quickly resolved by telling the crew to switch the instrumentation system to a backup power supply. This was not to be the case this time. The data appeared incoherent because it was trying to reflect simultaneous multiple system failures. My response of a possible instrumentation problem, however stated, was essentially only a ploy to put Kranz off for a few more seconds so that I could gain more insight into the problem.

But the *Apollo 13* failure had occurred so suddenly, so completely, with little warning, and affected so many spacecraft systems, that I was overwhelmed. No amount of training or mission experience could have prepared me to quickly "psych" out *this* problem. As I looked at my data and listened to the voice reports, *nothing* seemed to make sense. However, it soon dawned on me that diagnosing and dealing with the problem was exclusively my responsibility.

Reality set in when Tank 1's pressure dropped from 880 psia to about 400 psia within three minutes and two of the fuel cells failed to respond to troubleshooting. It wasn't yet evident, but the shock of the explosion of Tank 2 had cracked a line to Tank 1, causing it to leak. The three fuel cells were located on the shelf above the oxygen tanks. The shock had closed the gas feed valves to Fuel Cells 1 and 3 "starving" them to irreversible failure within a minute. Fuel Cell 2 was still operating, but the oxygen it required was rapidly leaking away. Two-thirds of the electrical power generation system was dead and we now had only three hours of life remaining. With one of the ship's two electrical power distribution systems shut down, many pieces of vital equipment were no longer being powered. My back room guys and I had to work backward from the symptoms to discover the root cause of the problem. As I experienced my first cascading multiple systems failure, I was discovering an example of the Domino Theory.

It was a most unwelcome example.

As the emergency developed, people spilled into the MOCR and the noise level rose, so Kranz admonished, "Let's everyone settle down. Let's not make matters worse by *guessin'*."

Eighteen minutes had passed, but I was oblivious to the passage of time.

With two fuel cells apparently off-line, I instructed Flight to have the crew begin an emergency power down in order to lighten the load on the single remaining fuel cell, which had become overloaded. Fred Haise, anticipating our next move, brought Entry Battery A on line to support sagging Main Bus A voltage. Our strategy in dealing with problems was to minimize the impact on the mission. Kranz cautioned me in clipped tones, "Let's make sure that we don't blow the whole mission."

None of us had accepted the possibility that the mission was already blown.

It was an incredibly tense time; a monster CSM failure had occurred and I had no quick answers. As EECOM, I felt very much on the spot; I could feel the chill of panic begin to well up – there was no one to whom I could turn, and I will admit that a fleeting thought of getting up and going home *did* cross my mind. Of course, that was not an option! I firmed my grip on the "security handles" of my console, and brought my emotions under control.

As it dawned on me that we'd lost two fuel cells and I didn't understand the reason why, I started thinking that we might need to get the crew into the Lunar Module, which was still powered down. We were fortunate that we could use the LM as a 'lifeboat.' Of course, if the accident had happened while Lovell and Haise were on the Moon, or after they had returned and set off for Earth, this option would not have been open to us. If *Apollo 8* had suffered such an accident its crew certainly would have perished within hours.

Dick Brown ("Brownie"), EPS flight controller

I stood up and called TELMU Bob Heselmeyer, my LM counterpart. He was only ten feet to my right, but I called him on his loop. "Bob, do you remember the lifeboat procedures that we started to develop on that sim when we went around the Moon and I got caught with the loss of cabin pressure? Did you guys ever work on those lifeboat procedures?" Bob was soloing for the first time as a TELMU. He stared at me, unresponsive, most likely overwhelmed by the unfolding mission events. With the realization that I was alone, I slowly sat back down in my chair, resuming my grip on both "security handles."

> *Gene Kranz had once said that the hardest part of being a flight controller was being "the last person in the decision chain." That occurred, Kranz said, when a problem was passed to a flight controller at the last minute and he had to solve it all by himself; if he made a mistake, he did it in front of the whole world.*

Good coordination back and forth with my back room specialists was vital to solving problems, and I had developed an excellent rapport with them. I set my SSR ECS George Bliss and ECS Larry Sheaks and EPS Dick Brown to work looking at their systems and offering troubleshooting suggestions to isolate the leak that was taking Oxygen Tank 1 down. We all had difficulty accepting that both oxygen tanks could be lost, let alone that two fuel cells were dead. What transpired was an interminable series of configuration checks and changes as I directed the flight crew in the futile attempt to isolate the leak and recover the fuel cells.

It was now 20 minutes since the failure. The last two minutes seemed like two hours.

I informed Kranz, "Flight, I've got a feeling that we've lost two fuel cells ... I hate to put it that way, but I don't know why we've lost them. It doesn't all tag up. And, it's *not* an instrumentation problem." This was the news the astronauts dreaded to hear. The possibility of a lunar landing was definitely gone because the flight mission rules precluded a lunar landing if one fuel cell was lost, let alone two – and a whole lot of other lost equipment. EPS Dick Brown and I believed that Fuel Cell 3 was a likely source of the oxygen leak and needed to be isolated, so I explained this to Kranz and recommended that both reactant feed valves of Fuel Cell 3 be closed. Confirmation of the loss of the lunar landing mission came to the crew when CapCom Lousma relayed this instruction to the crew. Fred Haise, knowing the consequences of closing the valves, repeated the request, followed by, "Did I hear you right?"

The nightmare was still in full force after 38 minutes.

We watched in disbelief as Oxygen Tank 1's pressure continued to decrease despite our efforts to isolate the leak. When Larry Sheaks, concern evident in his voice, told me that the Surge Tank pressure had also begun to drop, I immediately called Kranz "Flight, let's have the crew isolate the Surge Tank." This came on top of my call to power down as much CSM equipment as possible in order to save the battery that was supplementing the one remaining fuel cell. In effect, our strategy was to match the load to the output capability of the remaining fuel cell. By isolating the Surge Tank I seemed to be suggesting that we cut off another source of oxygen, albeit a small one at only 3.7 pounds, to the remaining fuel cell. Gene momentarily drew a blank and uttered another statement which would come back to haunt him later, "I don't understand that, Sy." I reminded him that whereas the main oxygen tanks were in the SM, the Surge Tank was in the CM and it supplied oxygen during the entry phase, and I wanted to save it. He may or may not have been familiar with the system, but he immediately appreciated the significance of the Surge Tank and gave the order. Later, I requested the isolation of the Repress Pack, since its three one-pound bottles could also serve as a backup supply of precious oxygen for entry. These actions were to preserve capability that the crew would need for re-entry at the mission's end.

Loss of the crew was never an option I considered.

Kranz's voice broke before mine did. He doesn't remember. But he was cool. He could amalgamate a lot of information at one time and keep the team focused. He knew what to do. He kept us calm. He kept things flowing by giving orders such as, "let's get the crew calling down configuration" . . . let's do this; let's do that.

Many years later, after listening to the audio tapes of my EECOM console intercom loops, it amazed me that I was able to maintain an outwardly calm demeanor throughout this episode. Certainly the training helped, but I can't help thinking of the influence of the protective emotional armor that I had developed from the childhood parental discipline, beatings, and inattention.

Now forty-six minutes into the emergency.

I controlled the tone of my voice so that it didn't betray my growing despair, and informed Kranz, "Flight, the pressure in O_2 Tank 1 has dropped all the way down to 297 [psia] and we'd better think about gettin' in the LM." In this battle, my plan was to keep retreating by drastically powering down the CSM's systems to lighten the load on the remaining fuel cell in order to stretch it out as long as the ebbing oxygen would allow. It became a race to activate the LM before the CSM died.

I suddenly realized that I had ceased swallowing.

I called back to my ECS guys with an urgent request for the remaining lifetime for O_2 Tank 1. A few minutes later, ECS George Bliss reported to me that at the current leak rate, Tank 1 would last 1 hour 54 minutes. By that time it would have fallen to 100 psia and the last fuel cell would die of starvation. After that, the batteries would be the only source of power in the CSM, but these had to be saved for entry. No sooner had I been informed of this time limit then Kranz asked me if I had any ideas for increasing the pressure in O_2 Tank 1. I said that this was not an option because we couldn't stop the leak, and I told him starkly, "Flight, we're going to hit 100 psia in an hour and fifty-four minutes."

That's the end ... right there."

I remember at one point, seemingly out of the blue, Kranz called down to the Real Time Computer Complex (RTCC) supervisor, asking, "How many machines do we have on line? I want you to bring another one up to run delogs," that is, to play the history data into computers and print it out so we could analyze it to trace the causes of the failures. After the mission, I asked, "Kranz, how on Earth during the midst of that chaos, or near-chaos, did you think to call up another machine to print out delogs of the data?" He replied, "I had it on a checklist on my console." The master that he was, the man had thought ahead and planned that if we had a catastrophe, one of the things he must do was to get the data printed out so it could be analyzed.

To an observer, the MOCR was eerily quiet throughout this critical period, but the intercom loops were alive with discussion of the growing emergency.

So there I sat, during the last hour of my shift, looking at telemetry data that indicated the most improbable multiple failures. I had two of the three fuel cell electrical power plants dead, one of two power distribution systems unpowered, a whole tank of oxygen gone, and the other one leaking. The CSM had less than two hours of life remaining and Kranz and CapCom Jack Lousma, who sat two feet to my left, were pressing me for recommendations! I overheard Craig Staresinich, who was a new trainee working in the SSR, say despairingly to no one in particular, "I don't know where to start."

Kranz realized that the best thing we could do at that point was to get off the consoles and let Glynn Lunney's Black Team come on and continue the work. Another Flight Director might have said, we're staying here until the bitter end, but such was Kranz's discipline that he announced, "We're off-duty, let's just hand it over to a fresh team and let them tackle the problem." All our counterparts on Lunney's team already knew what was going on because, as was the practice, the new team's flight controllers had come in an hour early to familiar themselves with the situation and the console logs, and we smoothly transitioned over.

> *Ironically, if I hadn't requested the extra cryo stir, the problem would have happened to the next EECOM eight hours later. I've heard it said: no good deed goes unpunished.*

I must admit, I was *very* relieved (no pun intended) to hand over to EECOM Clint Burton, even though it was a very hot potato. Over the next two hours, Clint oversaw the inevitable failure of all power and oxygen in the CSM as other members of the flight control team worked with the astronauts to activate the LM. Unfortunately, the Lunar Module did not have fuel cells, only batteries, and these were insufficient to supply full power throughout the loop around the Moon and the return to Earth, requiring a power down of most of the LM's systems. The consequence of the power down was that Lovell, Haise, and Swigert had to endure four days in a cabin environment comparable to the inside of a refrigerator.

With the CSM dead and frozen, there wasn't much for me to do, except come each day to sit at my console during my shift. "Tiger Teams" were formed to deal with the trajectory decisions, LM power management, and the procedures to re-power the CM for entry without prematurely using up the power in the three small Entry Batteries. In sustaining the CSM during LM activation, these batteries had been discharged to half-capacity. This would have been a major problem if Bob Legler, LM EPS, had not determined that the Entry Batteries could be recharged from the LM batteries, using in reverse, a wiring circuit that was originally meant to send CSM power to some LM equipment heaters during the Trans-Lunar Coast phase. That was one of the many "biggies" that saved the *Apollo 13* astronauts.

There were many heroes on the Apollo 13 *mission.*

John Aaron headed up the CM power-up Tiger Team and, as a "Power Broker," forged a set of procedures that allowed an absolute minimum amount of spacecraft equipment to be re-powered. To do so, a ground rule was agreed upon that redundancy would no longer be mandatory. The next few days were hectic as the procedures were developed, checked and rechecked in the simulator. Finally, with little time to spare, they were read up to the crew, who were anxious to have them because if the radio link had failed while they were waiting, they would have been doomed – there was only one chance to get it right.

The final high drama of the mission was about to begin.

I relinquished the console to John Aaron, and moved over to the right seat to help out as the astronauts worked through the power-up procedures to resurrect the dead and frozen CM. Fortunately, the batteries were fully functional (no, they weren't Die-Hards). However, they had to last through the fiery entry right through to splashdown. It was a seemingly unrelenting period of tension as John and his back room EPS Jim Kelly coordinated the activity.

Apollo 13 SM damage seen after its separation

We held our collective breath as each piece of essential equipment was turned on. Suddenly, Aaron called out on the Flight loop, "Where's that extra two amps coming from?" GNC reported that part of his backup guidance system was on. "Tell the crew to turn it off!" John ordered urgently. If the batteries drained, the craft and the crew were doomed. It was reminiscent of the nervousness that I'd felt during the *Apollo 11* descent to the lunar surface.

Finally, with the CM properly re-powered, the SM was jettisoned and the astronauts got their first view of the damage caused by the

Apollo 13 on the main chutes

explosion. Lovell reported in amazement, "There's one whole side of that spacecraft missing!" The photos they took were startling and sobering.

They had narrowly dodged a bullet.

Finally, an hour and a half before splashdown, the LM that had sustained the astronauts' lives for four long days was jettisoned too. "Farewell *Aquarius*, and we thank you," said CapCom Joe Kerwin. "She sure was a good ship," agreed Haise, who'd hoped to land her on the Moon. The CM and its crew of superb pilots, Lovell, Haise and Swigert, were now on their own, falling at 25,000 miles per hour towards a narrow corridor in the Earth's atmosphere: too steep and it would burn up; too shallow and it would bounce off back into space.

We watched the data become static as the CM entered the communications blackout, and we became anxious when this extended beyond the normal three minutes. Had they come in too shallow and skipped out of the atmosphere? After an agonizing six minutes, we heard the crew report that they were on the main chutes. RETRO Chuck Deiterich, told me later that he believed they had probably entered a little shallow, prolonging the blackout period. What a beautiful sight it was to see the Command Module on the main chutes and then splashing down in view of the recovery carrier. We were all so relieved and so very proud.

For us flight controllers, the mission was a success.

Apollo 13 Splashdown

How close did we cut it with the Lunar Module consumables? Although the LM had 104 hours of breathing oxygen remaining when it was jettisoned, it had only *10* hours of water and *13* hours of electrical power remaining.

It had been damned close.

Apollo 13 MOCR after crew on the carrier. I've already gone home

That awful hour of my shift, when I was nearly overwhelmed by the crisis, had badly shaken my confidence. Flight controllers were expected to be always on top of the situation, so I felt that I had failed, and that somehow it was my fault because I surely hadn't done all the things I could have done. After the crew was safely aboard the aircraft carrier *Iwo Jima* I felt no elation; I simply left the chaotic celebrating in the MOCR and went home.

The bad dreams began almost immediately.

I began having a wake-up dream about the *Apollo 13* incident. As the last dream before you wake up, it can be so real, and the one that you recall. My new wake-up dream had me reliving the oxygen tank explosion. It sounds like psychobabble, but every morning just before I awoke, I'd dream that I was sitting at my console, the tank blew up, and the fuel cells died, etc., and I went through all the steps as it really happened. *It was very real.* I was at a low ebb. It was as if I was being punished for failing to live up to the portrayal of the perfect flight controller as a "steely-eyed missileman" who was always on top of the situation.

Amazingly, after two weeks the dream took a different course. This time, as the stir began, I spotted the tank pressure climbing and I made the call to the Flight Director, "Flight, looks like the pressure's going up in O_2 Tank 2; it looks like the heaters have failed on. Have the crew pull the breakers on Panel 226." I followed up in rapid fire with other instructions. "No joy, Flight; it looks like we've got a fire in the tank." And then the tank blew up, which I reported accurately. I was making all the correct calls, fuel cells, everything. It was perfect. I felt great. After waking, I noted that my perfect actions had made no difference to the final outcome. With this realization, I never had the dream again. The mind is a wonderful thing.

So I wasn't a failure after all ... my subconscious told me so.

The Lunar Module flight controllers have always taken second stage to the CSM guys, most likely because the LM always performed its mission with near perfection. There has been little, if any, historical recognition of the heroic efforts of the LM flight controllers as they scrambled to direct and monitor the LM activation before *Odyssey* went dark, and then began to implement a LM power down procedure to conserve its precious consumables for the long journey home.

Actually, LM lifeboat procedures had already been developed a year before the *Apollo 13* disaster occurred. On April 25, 1969, during an *Apollo 10* simulation, the CSM sustained a hydrogen leak at 51:15 hours into the mission, which resulted in the loss of all three fuel cells. Unprepared for this kind of failure, the flight control team was unable to save the crew.

Eerily, the time that the sim guys had selected for the failure to occur was nearly the same as when Apollo 13 *was struck.*

Understandably, Glynn Lunney, who was the Flight Director during that sim, minimized the significance of the loss, saying it was an unrealistic multiple failure case.

Nevertheless, Jim Hannigan, the branch chief of the LM flight controllers was concerned. Few people are aware of the fact that after that sim he authorized a small Tiger Team to be formed to develop some "LM Lifeboat Cases" to be used in the event that all CSM power or some other critical function was lost. Jim gave the job to Don Puddy, then a TELCOM (and later a Skylab Flight Director), who assembled a small group of LM and CSM flight controllers who worked on developing the procedures all summer and fall of 1969, even during the flights of *Apollo 11* and *12*. Several cases were developed and "put on the shelf" for future use, since a decision was made not to include them in the onboard crew checklists. As fate would have it, those procedures were required on the very next mission and they were dusted off, hurriedly updated to reflect the actual situation and tested in the simulator. Since it is not mentioned in his book, *Lost Moon*, Jim Lovell probably was unaware of this critical planning activity.

Apollo 13 Spashdown party at the Hofbraugarten biergarten. Sy presents a commemorative plaque to Jack Swigert and Jim Lovell

Apollo 13: Nixon at MSC presenting the Medal of Freedom

There is a strong possibility that if it were not for those previously thought-out cases and procedures for using the LM in a manner for which it was not designed, the Apollo 13 *crew might have been lost.*

There were many celebrations by various sponsors up and

Apollo 13: Nixon at MSC wide crowd shot

Medal of Freedom certificate with Sy's name

down NASA Road 1, but the only one that mattered was our traditional "splashdown Debriefing" party with the crew in the Hofbraugarten biergarten. Once more, we flight controllers retrieved our mission mugs, filled them with beer, and toasted the crew and each other.

Astronaut John Young, the backup CDR, showed up at yet another post-mission celebration that was hosted by the *Apollo 13* flight crew. He announced that he had an audio tape that he wanted everyone to hear. It was a hastily prepared tape produced by editing music and lyrics with clips of things that were said on the control center intercom loops and by the crew in space. The result was a "roasting" of Kranz and me with clips like my, "We've got an instrumentation problem, Flight," followed by (and completely out of context) his, "I don't understand that, Sy." Those two phrases were repeated over and over interspersed with other funny stuff. What a hoot.

To this day, Young denies any knowledge or connection with the production of the tape.

The largest celebration took place at MSC. Thousands of employees gathered outdoors and witnessed President Nixon awarding the Presidential Medal of Freedom to the members of the *Apollo 13* Mission Operations Team for our heroic efforts in saving the astronauts. I treasure that award.

I've often been asked why there are no pictures of the EECOM console during *Apollo 13*, and indeed most other missions. The reason is simple: whenever there was a perceived commotion about the mission, the two photographers in the MOCR, one still and one motion picture, knew to immediately swing over to document the

Flight Director's console or the CapCom console. As the photographers were conscientiously getting shots of the crowd at the CapCom console (Jack Lousma and three or four other astronauts) during the *Apollo 13* crisis, I was hanging onto my console's "security handles" two feet to the right, working the problem. One of the photographers actually sat on the left edge of my console so as to steady himself. They didn't have a clue as to the details of what was going on, nor did the Public Affairs Officer at his console in the MOCR.

Apollo 13 crowd around CapCom console. L-R: Deke Slayton, Ken Mattingly, Vance Brand, CapCom Jack Lousma, John Young. The photographer had to be sitting on my console

The Cortwright Committee was formed and convened to conduct an investigation into the cause of the *Apollo 13* incident and to make recommendations. In the expectation of being called to testify, I had the voice recording station in the Mission Control Center immediately make me a tape of all the intercom loops that I had activated on the EECOM console, covering the period of time from just before the tank explosion to the ultimate loss of all oxygen and power, approximately three hours later. However, much to my relief I was not called to testify. I dropped the two slim boxes containing the open-reel tapes into a drawer of my desk where they remained undisturbed for 18 years. (Yeah, so I don't clear my desk out very often; who does?)

One day in 1988, I received a phone call from a writer who introduced himself as Charles Murray. I had no idea who he was, but that was my failing, because he had a Ph.D. in Political Science from MIT and was a best-selling author. He told me that he was in the final stages of a five-year project to write a book about our space program period up to the landing of men on the Moon, and that every time the subject of *Apollo 13* arose, my name came up. He made a decision that he'd better talk with me before wrapping up the project. I invited him up to my office and we talked for about an hour before I recalled the existence of the tapes. I agreed to let him have them dubbed to standard cassette tapes. When he returned the next day with the tapes, I asked him if he had listened to them and he replied with wide eyes, "Holy shit! That was really something!" When I told him that I had listened to the tapes only once in 18 years, he said I should listen to them again. I have, but only once more. He then wrote an additional Chapter 27, based on those tapes*, for his book *Apollo: The Race To The Moon*. Although there were only two print runs, the book sold out and is now highly sought after in secondhand stores and on the internet.

*Those tapes containing the voices on the intercom loops on my EECOM console are included in the book CD-ROM. They cover the four hours from the explosion of O_2 Tank 2 to the depletion of the remaining, leaking O_2 Tank 1 that resulted in the loss of all power to the CSM.

— Chapter 46 —
Apollo 13: Trail to the Movie

In the 1990s, disappointingly, most people either never knew or had forgotten the details of our manned lunar exploration missions, let alone the high drama of the *Apollo 13* deep space rescue, and some crazies even argued that it had all been faked. Fortunately, much of the mission detail has been preserved for the public in books, TV documentaries, a motion picture and on a CD-ROM.

Henry S. F. Cooper, of the *New Yorker* magazine, was the first to document *Apollo 13* in his 1972 book *Thirteen: The Flight That Failed*. There were also a succession of space experience books by former astronauts that flew successful missions, but few of them fired the public's imagination.

Universal Studios filmed a program for the ABC TV Network entitled *Houston, We've Got A Problem*, which aired March 2, 1974. NASA, naïvely thinking that the production would be an authentic dramatization of the *Apollo 13* mission, allowed them to set up in the 3rd floor MOCR, which by that time was not in use. In actuality, the show was pure fiction, a soap opera framed by a thin halo of actual space radio transmissions and stock film footage. I was portrayed by Steve Franken as "Shimon Levin," a Jewish drug addict who cheated on his wife and suffered a rabbi brother who was determined to reform him. Robert Culp, playing RETRO, collapsed in the hallway and suffered a heart attack brought on by the stress of the dire situation. I gave this farce a wide berth, but many of my fellow flight controllers participated as extras. Frequently thereafter, I was asked if that was really the way *Apollo 13* happened.

For the next 15 years, there was a dearth of interesting publications about our country's space accomplishments.

In 1989, Simon & Schuster published *Apollo: The Race to the Moon*, written by noted author Charles Murray and his wife, Catherine Cox. Despite only two printings, the book became accepted as *the* definitive book describing the development of our nation's space program to the ultimate goal of landing a human on the Moon.

Late in 1992, Jim Lovell began to actively work on his autobiography in collaboration with Jeffrey Kluger. Entitled *Lost Moon*, much of the book is devoted to his final space mission, *Apollo 13*. Later, I received a call from Jeff asking me for historical assistance. I challenged him that he was writing another "astronaut book" in which the story is dominated by the activities of the flight crew. Jeff protested that Lovell fully recognized the role of mission control in saving their lives and wanted to tell the whole story. On that basis Jeff was able to elicit the necessary cooperation to pen an artful, balanced account.

The first Moon landing has produced a plethora of books written by astronauts and others about their lives and space exploits. One book in particular stands out among the best: in 1994, after eight years on a labor of love, Andy Chaikin published his space tome, *A Man On The Moon*. It is an authoritative, in-depth recounting of the experiences of all the lunar astronauts – their fears and the exhilaration of their triumphs. The book became the main basis for Tom Hank's Emmy-winning HBO mini-series *From the Earth to the Moon*.

On July 20, 1994, on the 25th anniversary of our first manned lunar landing, PBS aired the epic documentary of the ill-fated mission, entitled, *Apollo 13 – To The Edge And Back*.

That same year Fred Schoeller, owner of *Arts & Letters*, approached me to collaborate on a CD-ROM about *Apollo 13*. The result, a year later, was an archival CD-ROM entitled, *Apollo 13: A Race Against Time*. Fred grieved for the golden days of Apollo. I convinced him to drive down from Dallas, Texas, so I could give him and his CD-ROM collaborator Mark Caterina a tour around JSC, especially the old MOCR, which is now a national historic site. I walked with them among the old consoles, now configured as they were during *Apollo 11* and explained who sat where and who did what. When we finished, Mark preceded me out into the hallway and I held the door open for Fred, but he lingered just inside the door for a last look at the scene, frozen in a pose with his hands clasped at his chest just under his chin as if in prayer, his face betraying an expression of sorrow. *He was grieving*. To this day, memory of that poignant scene still nearly brings me to tears.

On Monday, April 11, 1994, I received a phone call from Gene Kranz' secretary to be at Pe-te's (pronounced Pee-Tee's) after work for a beer and BBQ meeting with Jim Lovell, who wanted to discuss a project to make a movie based on his book. Pe-te's, was a ramshackle building across the highway from Ellington Field, cobbled from a 1950s gas station. It had a menu featuring Cajun food and so-so BBQ. As I walked into the large back dining room, I was surprised to see Gene Kranz, Jerry Bostick, Gerry Griffin and John Llewellyn in animated conversation with Jim Lovell, Ron Howard, Tom Hanks and Tom Pollock, a producer. It wasn't long before the mission control anecdotes were flowing freely and, as the beers worked their magic, Llewellyn and Kranz commenced to out-mission con-

L-R: Mark Caterina, Fred Schoeller, Sy Liebergot at restaurant in Kemah Texas. Fred, owner of Arts & Letters, produced the Apollo 13 A Race Against Time CD-ROM. Mark was a collaborator and Sy the Technical Advisor

trol war story each other.

This inconspicuous gathering launched the process of creating Ron Howard's wonderful movie, *Apollo 13*. Ron was determined to portray the story as authentically as possible and he had a near-exact studio replica of the MOCR constructed. Gerry Griffin, an *Apollo 13* Flight Director and a movie technical advisor, told me, "Sy, it was eerie, the first time that I walked into the studio MOCR, I thought that I was actually there." Fellow flight controller Bob Legler and I spent long hours supplying the movie company's contractors, such as the property manager, a myriad of authentic details ranging from the documents we had at the consoles to the identification badges we wore.

One day I received a call from Clint Howard who informed me that he would portray me in the movie, and he wanted to make sure that he really got "into character." I asked him if he had read either Cooper or Murray and Cox's books, if he had viewed the PBS *Apollo 13* documentary, and listened to my console audio tapes; he had indeed. I was impressed and we began a series of lengthy telephone conversations. We became good friends as a result of the movie involvement; to this day Clint will occasionally ask me if his performance was okay. The *Apollo 13* movie provided another bonus: it allowed me to know and become friendly with his brother Ron, and Tom Hanks, who played Jim Lovell. As the filming drew near, Clint asked me if it would be all right to use my name in the movie, and I was taken aback that they might not. I replied, "Clint, I would be honored if you guys used my name in the movie. I would be disappointed if Ed Harris, who brilliantly played Gene Kranz, at a moment of high tension, called me 'Lou'."

We both had a good laugh over that, and so "Sy," it was.

On June 12, 1995, Ron Howard, ever appreciative, held a private VIP reception at a nearby resort for many of us space veterans and the stars of the soon-to-be released movie. The surprise came when nearly two hundred of us boarded five comfortable buses to an unknown location, which turned out to be a theater where we were treated to a preview showing. Ron used the opportunity to thank everyone for his or her participation and help in making the movie. Tom Hanks, sporting a new beard, followed similarly, but added, "How do you like my Sy Liebergot look alike beard?"

Shortly thereafter, on June 30, 1995, the movie *Apollo 13* was released to theaters around the United States. The movie had sell-out showings and audiences applauded in the theaters, which seemed to demonstrate that Americans still loved their heroes and have great pride in their country's space accomplishments. Even young people, who were born long after Apollo, displayed pride and enthusiasm.

Thank you, Ron Howard.

Clint Howard and Sy at Apollo 13 30th Anniversary fund-raiser gala for the Astronaut Scholarship Fund

2000: Apollo 13 30th Anniversary astronaut scholarship fund-raiser in Santa Monica, California. L-R: Melanie Howard, Clint Howard, Craig Liebergot, Sy Liebergot

1995: Apollo 13 25th Anniversary astronaut scholarship fund-raiser in Southern California. L-R: Craig Liebergot, Ron Howard, Sy Liebergot

1995: Private reception and preview showing of Apollo 13 movie.
L-R: Sy Liebergot, Craig Liebergot, Tom Hanks with his "look alike" beard

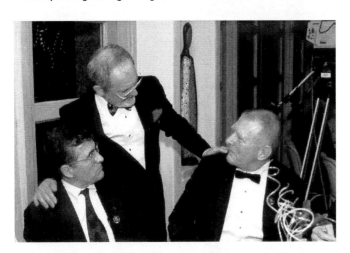

1995: Apollo 13 25th Anniversary astronaut scholarship
fund-raiser in Southern California.
L-R: Jack Schmitt, Sy Liebergot, Gene Kranz

— Chapter 47 —
Apollo 14: Bruce McCandless

Apollo 14, launched on January 31, 1971, was assigned the *Apollo 13* landing site at Fra Mauro. Aboard the CSM were CDR Al (Big Al) Shepard, LMP Ed Mitchell and CMP Stu (Smokey) Roosa, a former "smoke jumper" in the US Forest Service who parachuted into forest wildfires in order to put them out. Al couldn't wait to practice his golf swing on the lunar surface and Ed looked forward to trying to contact people by ESP from deep space.

After my *Apollo 13* experience, I was assigned as Lead EECOM for *Apollo 14*. I was especially nervous about this mission because it carried the modifications to the CSM recommended by the Cortwright Committee that investigated the *Apollo 13* accident. The major modifications were an additional cryogenic tank of oxygen, a 400 amp-hour LM Descent Battery to back up the loss of all three fuel cells, and plastic water bags to store potable water, which had frozen after the CM went dark on *Apollo 13* and resulted in crew dehydration. The extra oxygen tank was located on the far side of the SM from the other two, as an additional precaution.

My general nervousness was not equal to that felt by LM Control flight controller Dick Thorson. In lunar orbit, just four hours before Powered Descent Initiation (PDI), Thorson reported that an Abort command had been set in the LM computer, even though the crew had not actuated the Abort switch. If the fault had not been detected, the LM would have aborted immediately after the Descent Engine fired. The Abort command was procedurally reset, and each time it reoccurred it was reset by tapping on the panel near the Abort switch; the most likely culprit was a "solder ball" or a "wire whisker" floating inside the switch module. Within two hours, a software "patch" was written, read up and entered into the computer by Mitchell, to instruct the computer to ignore the switch. The spurious Abort command was not visible to the two astronauts, so Thorson's diligence had saved the landing.

Astronaut Bruce McCandless was some flight controllers' favorite whipping boy; he seemed to always screw up something in the control center during sims or missions when he sat at the CapCom console.

During one shift, we EECOMs had the crew perform a procedure called Quantity Balancing on the three cryogenic oxygen tanks. The O_2 Tanks 1 and 2 were plumbed together and fed simultaneously, O_2 Tank 3 was the new tank and could be isolated from the other two tanks and the fuel cells. Although in normal operation all three tanks fed the cabin and the fuel cells simultaneously, and depleted at an equal rate, sometimes one depleted more rapidly than the others, creating a quantity imbalance between them. The mission rules restricted this imbalance to no more than 2.5

percent, and in this case the imbalance was approaching 5 percent, so a special corrective procedure was called for. This simply required the crew to turn off the automatic pressure control heaters in the tank with the lowest quantity reading, thereby forcing the tanks with the greater quantity to feed more oxygen until their quantities closely matched that of the other tank. However, on Mission Day 8, while having the crew perform the tank balancing procedure, the fact that the oxygen tank quantities were nearly at 50 percent imposed a high temperature restriction of +350 degrees F on the internal tank temperature, and it was not allowed to rise above that value. This temperature was displayed on my EECOM ECS screen, and realizing that the temperature would overshoot a few degrees after the heaters were turned off, I told the Flight Director and McCandless that I would "lead" the overshoot by calling the heaters off at 325 degrees. The command I wanted directed to the crew was simple: O_2 Tank 3 heaters to OFF, O_2 Tank 1 and 2 Heaters to AUTO, and CapCom McCandless duly voiced the instructions up to the spacecraft.

Bruce, had meanwhile called up my ECS display on his console screen to follow along with the procedure. He apparently became confused, and seeing the temperature pass 325 degrees, made a call to the crew without authorization that repeated the switch configuration they had just been asked to select and already confirmed as completed. I yelled at Bruce, "No, no, they've already done that!" Too late. Big Al called down, his voice dripping with acid, "Roger, Houston, We've got Oxygen Tank Heater number 3, OFF. O-F-F! We have Oxygen Tank Heaters number 1 and number 2, AUTO. A-U-T-O!" I was mortified and laid my head down on the console desktop, figuring that Big Al didn't think much of the EECOM now.

A most practical joke:

There was a block of 48 small, cube-shaped gray lockers in the hall behind the MOCR. Other than the numbered label riveted on the door, they were indistinguishable. They were secured with a combination lock. One day, upon arriving for a day of *Apollo 14* simulations, as I retrieved my headset from the EECOM locker I noticed a bright yellow plastic label embossed with the words "CapCom Locker" on one of the locker doors. I immediately knew it had been placed there by Bruce. Laughing to myself, I thought, "the sonofagun can't remember which was his headset locker!" Later, I called Big Al (Al Shepard) and proposed a plan to gig Bruce. He loved it. So, with his approval I made 47 additional identical yellow labels that read CapCom Locker, as well as others that specified directions to the CapCom console.

On mission day, I arrived at the Control Center before Bruce, and after applying the other forty seven labels to the lockers, I affixed the others in sequence along the walls announcing: "This way to the CapCom console"; "You're getting nearer to the CapCom console"; "Here's the CapCom console"; "You're here"; as well as "Right" and "Left" on each armrest of his chair, "Front" and "Back," so that he wouldn't risk getting in the chair backwards, and even "CapCom TV Monitor." Bruce was not amused. He was fuming as he took his seat at his console which was only a couple of feet to my left. Of course, he knew that I was the perpetrator. Anyway, we launched and after a couple of Earth orbits, the Trans-Lunar Injection burn was performed by the S-IVB and we were on the way to the Moon. After TD&E, when there was a lull in the activities, Big Al called down with, "Hey Bruce, did you have any trouble finding your headset today?" Damn, that was fine! Everyone (I had made sure that everyone was in on the gag) burst out laughing. Bruce fumed even more. We EECOMs could be pretty ruthless. We didn't show Bruce much respect because he had a reputation among some of us flight controllers for goofing up. I believe he earned seven "DFMs" during the mission.

Ultimately, Bruce would be instrumental in the design and development of the Manned Maneuvering Unit (MMU), and on Space Shuttle *Challenger*, in February, 1984, in the first MMU checkout, he made the first untethered space walk. The pictures of him free floating against the black of space and with the Earth as a backdrop are breathtaking.

In the end, Bruce showed us all.

— Chapter 48 —
Apollo 15: Some Flight Controller Examples

Apollo 15, launched on July 26, 1971, carrying Commander Dave Scott, Lunar Module Pilot Jim Irvin, and Command Module Pilot Al Worden. We flight controllers called Worden "Little Al," in contrast to Al Shepard's "Big Al." The mission featured the introduction of the $40 million Lunar Rover Vehicle and the first televised lunar liftoff using the parked Rover's camera remotely controlled by INCO flight controller Ed Fendell. As a result of this innovation, Ed was dubbed "Captain Video" by the news media.

Forty-six minutes and 7,460 miles had elapsed since the TLI burn when Dave Scott made an alarming report, "… we've noticed that the SPS Thrust light on the EMS is now on and all the switches [the Delta-V Thrust A & B switches] are off."

The Entry Monitor System (EMS) provided guidance and control information for the crew during re-entry into the Earth's atmosphere. Additionally, it included a display to monitor the progress of engine burns – displaying the amount of velocity to be gained or lost (the Delta-V) – and a light to indicate that the SPS engine valves had been opened and that the engine should therefore be firing. It was this light which was now illuminated, though the engine was clearly not on. This problem was serious in that unsuccessful resolution could preclude the Lunar Orbit Insertion (LOI) burn and the lunar landing mission.

Although the spacecraft was still three days from the Moon, in the control center the flight control team and entire engineering community swung into high gear to address the errant light. The Gold Team GNC flight controller, Gary Coen, was the first to deal with the problem. His first task was to formulate a troubleshooting plan. Unchar-acteristically, Chris Kraft pressed Coen to request that the crew "tap on the panel" near the switches and light in question.

"Tapping on the panel" was also employed during Apollo 14 *when, in lunar orbit, an Abort command was set in the Lunar Module's computer, although the switch had not been depressed.*

INCO Ed Fendell's control of Lunar Rover Vehicle TV camera for Apollo 17 LM ascent

Gary politely told The Boss that he would not recommend that action (at least, not until he had a coordinated troubleshooting plan and test procedures). There were three possibilities: a short to ground in the light circuit; a short to ground that would light the SPS prematurely when the Delta-V switches were actuated; or a possibility of losing a redundant bank of engine valves. Ironically, part of the careful testing led the Maroon Team GNC on the next shift to safely request of the crew: "as a final check as to what's happening in that switch, we'd like you to tap around the Delta-V Thrust switches a bit. See if the light comes on." After six hours of careful testing, the problem was isolated to a set of contacts in the Delta-V switch that powered the light. Phew, the best possibility was the problem, and we would be "Go" for lunar orbit insertion.

It was at times like that I was glad I was EECOM rather than GNC.

In 1968, Milt Windler and Pete Frank became Flight Directors. Milt was transferred from the Landing and Recovery Division (LRD) and Pete was from the Mission Planning and Analysis Division (MPAD), both of which were in the Flight Operations Directorate, of which Chris Kraft was the Director. Their selection was a bit of surprise as neither man was a flight controller. The common belief at the time was that both were selected as part of an experiment by Chris designed to demonstrate that anyone (that he selected) could be a Flight Director; the implication that there was no one qualified among the current experienced flight controllers hung in the air. If Chris thought this to be the case, I don't believe he was entirely correct. The loyal, enthusiastic, and often brash Manfred "Dutch" von Ehrenfried, who had been with Kraft since Mercury, felt that he had been passed over for Flight Director and after serving his stint as Guidance Officer on *Apollo 8*, left the Manned Spacecraft Center, never to return. Milt, team color Maroon, began as a shift Flight Director on *Apollo 8*. Pete, his team color Orange, followed on *Apollo 9*.

Apollo 15 was my seventh mission as an EECOM, but it was my first time working with Milt. The changes made to the CSM for *Apollo 14* to prevent a recurrence of the *Apollo 13* nightmare had proved out, so I was looking forward to a no-sweat enjoyable mission. However, I knew that during any mission, something always occurred to keep me on my toes. To pervert an old adage, there are three things that made for a successful space mission: preparation, preparation, and preparation.

A good example of this happened on the third day out (at about 61 hours into the mission) when the potable water chlorination port leaked. While having their final meal of the day, the crew set about to complete some of the tasks in their pre-sleep checklist. One of these tasks was to chlorinate their potable water supply. In the pre-sleep checklist, the potable water tank was chlorinated to prevent bacterial growth within it. During the water chlorination procedure, a leak occurred, and an ever-growing globe of escaping water formed around the chlorination port. Dave Scott called down, "Okay, we're just getting ready to do some chlorination here, and we find we've got a leak around the chlorine port – with a cap on it – seems to be leaking water. And you might take a look at that real quick and see if you can come up with any ideas on the thing. The cap is on and Jim was just getting ready to take the cap off and noticed a little water; and, in trying to clean it up, it seems like we're accumulating a fair size – fair amount of water right now, right around the cap."

Astronaut and scientist Karl Henize asked, "Can you give us an estimate of how many drips per second it is?" Crew member Jim Irwin reported, "Yes, it's a pretty good flow right now. Drips per second? That's hard to measure; it's a whole ball of water right around that valve right now." From his Earth-bound perspective, Henize did not transform his thinking to onboard the spacecraft to realize that drips don't occur in zero-g, except where pressure is causing a jet to stream out; instead, surface tension forms the escaping water into an ever-growing globe around the chlorination port.

In space, small problems can turn into very big ones quickly, and this leak was threatening to do just that. The crew didn't yet know where the water was coming from. It could have been something as simple as a part that required tightening, but if it was a cracked pipe this would present a much more intractable problem. I quickly began working with ECS George Conway to find a way to take the pressure off the tank and slow the leak.

The chlorination port had leaked pre-launch and was reported by astronaut Karl Henize during his pre-launch cockpit activities. Now serving as CapCom, he told the crew of the earlier leak, and advised that when he had replaced the cap, the flow had reduced to a slow drip. Scott replied, "Oh, this is a big one, Karl, and the cap is on tight, and it – you can almost feel something flowing beneath the cap." Now with urgency in his voice, he followed with, "Got any suggestions yet? We need to isolate this thing pretty quick."

The crew did not have the knowledge on hand to deal with the leak problem and had no idea where the water was coming from, so they had to fall back on the considerable knowledge base of Mission Control and their support. Flight Director Milt Windler began winging suggestions to me. One such was telling the crew to dump water overboard from the tank, with which I was not ready to agree because it would affect PTC.

The combined spacecraft of the LM and CSM was slowly rotating around its long axis in a barbecue rotisserie fashion to assure even temperature over the whole structure. It was called Passive Thermal Control, more commonly PTC. It was tricky to set up, and venting water overboard would have undone the carefully set up spacecraft control mode.

Without realizing it, Windler was unnecessarily escalating the urgency of the problem and was interrupting my orderly troubleshooting discussion with Conway, who was now also telling me to have the crew dump water overboard. Bill

Moon, soon to be an EECOM, was working in SPAN and reminded me that an engineering record, called SPAN Chit #29, had been written pre-launch with the corrective procedure for the leaking chlorination port which simply read in part: *"put tool number 3 in the tool W ratchet, and insert tool 3 in the hex opening in the chlorine injector port and tighten a quarter turn."* I had it buried somewhere in the paperwork on my console, so Bill ran a copy of the chit out to me while I was still putting off more off-the-wall suggestions from Windler. After reviewing the procedure, I briefed Windler and Henize on the simple procedure and firmly suggested that it be immediately voiced up to the crew. Windler agreed, but he also wanted the crew to begin off-loading the water tank using their water gun into any available containers, to which I acquiesced.

Scott was appreciative, and quickly executed the procedure and reported, "Okay, Houston. It looks like that did it. Nice to have the quick response you guys have down there."

A few minutes later, Scott said with mock seriousness in his voice, "By the way Karl, we did a calculation on our slide rule and calculate that the leak was at least 3,000 drips per minute."

The crew should have been briefed in advance about the pre-launch leakage and the corrective action to be taken read up to them. If this had been done, their rapid action would have stemmed the leak before a large ball of water could form in the spacecraft. Even with the corrective procedures on the ground it still took 15 minutes to run it around the management track again, albeit this is pretty darn fast. On the basis that it was a "simple" fix, "Management" had decided not to bother the crew with this pre-knowledge in the form of an added written crew procedure, and it was decided to tell them only if the leak occurred.

In my opinion, a poor decision.

Scott later said of the incident, "Sy is correct, they should have told us before the launch. Which is another philosophical area to discuss – just how much should be sent up to the crew – three schools of thought (1) don't worry them; (2) don't worry them yet; and (3) tell them everything all the time. We favored the latter – give us the bad news first, the good news can wait – big boys need to learn how to deal with the bad news. But of course, don't complicate things with unnecessary details [which this was not; because of its potential consequences]."

Once again, "Let's not bother the crew."

(The actual on-loop conversations dealing with this episode can be heard on the CD-ROM included with this book. Heard are Flight Director Milt Windler, CapCom Karl Henize, CDR Dave Scott, Sy Liebergot and my back room ECS George Conway and Bill Moon in the SPAN Room. This serves an example of the pressures brought to bear on a flight controller to take action in an undisciplined manner.)

As a precaution after the *Apollo 1* fire, although the crew breathed pure oxygen in their pressure suits, the CSM was launched with a cabin atmosphere of 60/40 percent oxygen / nitrogen. The LM was also launched with a mixed gas cabin atmosphere, despite the fact that nearly all of its systems were unpowered and a blanket of inert nitrogen inside the Spacecraft / LM Adapter (SLA) surrounded it. The LM overhead hatch valve was left open, which allowed the LM's cabin to vent to vacuum during ascent.

Upon achieving orbit, the crew removed their helmets and gloves and began a series of planned CM cabin purges that brought the cabin to 80 percent oxygen by the TLI burn. After Transposition, Docking and Extraction (TD&E) of the LM from the S-IVB stage, the LM cabin was repressurized and equalized with the CSM by flowing CSM cabin air, essentially oxygen, through the CM Forward Hatch Equalization valve, through the connecting tunnel, and finally into the LM. The crew entered the tunnel for a quick checkout of the docking latches, installed two electrical umbilicals and then the CM hatch was reinstalled.

The presence of nitrogen was a medical concern, because if the astronauts were to avoid the "bends" when lowering the pressure in their suits for an EVA on the lunar surface, they would have to undergo a lengthy period of pre-breathing on pure oxygen. One of the planned crew activities on the way to the Moon was to bring the oxygen content and pressure of the atmospheres in both vehicles to at least 95 percent before undocking in lunar orbit. The required oxygen concentration in both spacecraft was finally achieved about halfway to the Moon, by venting some of the LM cabin pressure to space and replenishing it with pure oxygen from the CSM supply. The crew determined the necessary pressure by reading a delta pressure gauge in the CM forward hatch. This procedure was called "LM Enrichment." The pressure gauge reading was not telemetered to the ground, so if we wanted to be sure of the procedure, we had to ask the crew.

Apollo 15 was on its way to the Moon and things were very quiet when TELMU Merlin Merritt called me on my (intercom) loop to remind me that an hour had passed since the time that the crew was to perform the LM Enrichment procedure. Though coordination with the flight control team was not required, prior crews had coordinated the procedure with the EECOM. Since we could be an hour behind in the timeline, the upcoming ingress into the LM would be delayed and the concomitant TV show to the networks.

I asked Flight Director Milt Windler to inquire of the crew if they had performed the planned LM Enrichment yet. It was common practice that if a Flight Director required an explanation of a request that he did not understand, he would call you to his console and have you explain it off-loop, and then he would respond on the intercom loop so that everyone would be aware of his order. This ensured that the Flight Director loop was not tied up with a lengthy conversation. It was professional. By Windler's response, I could tell he did not know what I was talking about, but instead of asking me to explain off the loop, he began to question me in a manner designed to provide him the information without revealing that he hadn't known.

So, on-loop, I explained in detail what LM Enrichment was, and that we couldn't monitor the pressure gauge on the ground. The LM TELMU, Merlin Merritt, went through the same explanation and recommended that we ask the crew. Flight Surgeon Dr. John Ziegleschmid concurred. "We ought to ask the crew," prompted CapCom Dick Gordon. Finally, after *a cumulative hour* of this "justification" Windler gruffly consented, "Well, all right, but if you wanted to know, you should have put a 'Verify' in the flight plan!" Karl Henize, the prime CapCom promptly made the call, "*15*, we're anxious to know here if you have checked the LM / CM Delta-P." Dave Scott, now in the midst of preparations for the scheduled TV show, replied, "Not yet, Karl. We will though."

The reticence to ask the crew simple questions had always been a "hot button" issue for me because if we flight controllers needed information, then we needed information, and it was as simple as that. Furious, I stood up, turned to face Windler, and said hotly (off-loop, of course), "Windler, with an attitude like that, one of these days you're going to hurt someone!" His response was to lunge at me over his console muttering some epithet. I resumed my console, still seething. This reticence prevails today on the Shuttle and the International Space Station in the form of "Oh, let's not bother the crew with that." Hours, sometimes weeks, are spent on the ground on research and debate, when a simple question would provide the answer.

Some lessons are evidently never learned.

(The actual on-loop conversations dealing with this episode can be heard on the CD-ROM included with this book. Heard are Flight Director Milt Windler, CapComs Dick Gordon and Karl Henize, TELMU Merlin Merritt, Flight Surgeon Dr. John Ziegleschmid, EECOM Sy Liebergot and my back room ECS George Conway and EPS Charlie Jones. This serves a prime example of the interaction of flight controllers and the rapid exchange of information that occurred.)

Sadly, this was to be was my last mission of the Apollo Lunar program as an EECOM, because I made way for Bill Moon to be an EECOM on the final two missions.

Apollo 15: Sy Liebergot and Bill Moon at EECOM console

— Chapter 49 —
Apollo 16: Bodily Function

The fifth lunar landing mission began with the typical agonizingly slow liftoff of the huge Saturn V rocket on April 16, 1972. Aboard were Commander John Young, Lunar Module Pilot Charlie Duke, and Command Module Pilot Ken Mattingly.

This was my first mission away from the EECOM console – I was now working in the SPAN in the Systems position supporting the new EECOM Bill Moon.

This ambitious mission had a unique occurrence when a biological discomfort reared its ugly head, afflicting John Young on the lunar surface.

During an especially arduous *Apollo 15* EVA, the flight surgeons determined that both Dave Scott and Jim Irwin had suffered medical complications due to an electrolyte imbalance. In true "scientific fashion," after the manned space programs of Mercury, Gemini, and now Apollo, the medical experts spiked the crew's powdered orange juice with copious amounts of potassium. The result was a real-time commentary by John Young (who was unaware he had a "hot mic") while chatting with Charlie Duke in the LM on the lunar surface:

Young: I have the farts, again. I got them again, Charlie. I don't know what the hell gives them to me. Certainly not … I think it's acid stomach. I really do.

Duke: It probably is.

Young: (laughing) I mean, I haven't eaten this much citrus fruit in 20 years! And I'll tell you one thing, in another 12 fucking days, I ain't never eating any more. And if they offer to supplement me potassium with my breakfast, I'm going to throw up! I like an occasional orange. Really do. (laughs) But I'll be durned if I'm going to be buried in oranges.

There was no further mention of Young's problem during the EVAs. However, it should be noted that once inside the space suit the astronaut had only about 1.5 cubic feet of free volume, so any biological impertinence belonged solely to the producer.

— Chapter 50 —
Skylab: MSFC Steps to the Fore

The Skylab Earth-orbit mission lasted about a year, from May 14, 1973 to February 8, 1974. For us flight controllers, it was a totally different type of mission than a lunar flight.

Why Skylab? When Congress cut short the flights to the Moon by canceling *Apollo 18*, *19*, and *20*, it was really a false saving because the boosters and spacecraft were already built and paid for. Some people argued that because NASA had been created for a large project that it had accomplished, it ought to cease to exist. But it had grown to be a large and powerful bureaucracy, and it argued for another large follow-on project. During the 1960s, in parallel with the effort to place a man on the Moon, plans had been drawn up for a space station. Initially conceived as an adjunct to Apollo, Skylab became Apollo's successor.

As America's first space station, *Skylab*'s primary objective was to prove that humans could live and work in space for extended periods, but it would also be equipped to expand our knowledge of solar astronomy by making unique observations from above the Earth's atmosphere. It put some of the otherwise surplus Saturn and Apollo equipment to use. A "dry" third stage (S-IVB) of the Saturn V rocket was completely outfitted as an Orbital Workshop (OWS) by the Marshall Space Flight Center (MSFC) in Huntsville, Alabama, which thereby put its foot securely through the door of life support systems design, up to that time, the exclusive purview of MSC. The plan called for three crews to visit Skylab for periods of several months, flying up and returning to Earth in Apollo CSM spacecraft.

Disaster struck the unmanned *Skylab 1* during its launch on May 14, 1973. Part of the meteoroid shield that covered the Workshop was torn off, tak-

Skylab MOCR Identity Badge

ing with it one of the two large solar panel wings, and fouling the other so that it could not deploy. MSFC had decided not to subject the shield to aerodynamic stress tests, and instead qualified it "by analysis." Because the shield doubled as thermal insulation, once on orbit, Skylab found itself hot and underpowered. It was almost a total loss, but with no backup available NASA had to come up with a way to fix the crippled spacecraft.

The launch of the first crew was delayed until a thermal shield could be developed and tools manifested to free the jammed solar panel wing. Skylab's Apollo Telescope Mount (ATM) had its own set of solar panels that fed sixteen Charger Battery Regulator Modules (CBRM), so the station had some power. By deploying the parasol-type sun shield through a scientific airlock in the OWS, and later releasing the jammed solar array wing during an EVA, Pete Conrad's crew was able to accomplish its planned 28-day mission. The second crew went out and erected another sun shield over the first, using a V-shaped twin-pole device. Between them, these rescue activities meant that the rest of the Skylab Program was able to proceed much as planned.

The Skylab EECOM was called EGIL (pronounced EAGLE), the acronym being drawn from the words **E**lectrical, **G**eneral **I**nstru-mentation & **L**ife Support, which I thought was quite a reach for a call sign. EGIL had an even bigger job than the CSM EECOM: in addition to the life support systems, he had many more systems data to monitor and react to, and he could control some of the Skylab systems by commanding from his console. It was an enormous job of data and people management. Charlie Dumis erected a large brass eagle on a walnut wood base and we placed it on top of the EGIL console, which was now a three-bay, i.e., there were three CRT monitors, that displayed more than three times as much data as Apollo. There were also many more of the small hot "event" lights above and beside the CRTs as we had for Apollo. After *Apollo 15*, I was assigned to train as an EGIL, and my SSR staffing was increased to five specialists, in order to provide the necessary degree of support.

Near chaos reigned in the Mission Control Center as the engineering and flight control communities geared up to attempt to keep the Skylab alive during the ten

Skylab EGIL Sy Liebergot at EGIL console

Skylab 1974
Bronze Team
FD: Chuck (Skinny) Lewis
EGIL: Sy Liebergot

Skylab EGIL Sy Liebergot explaining to Flight Director Chuck Lewis

days that it took to prepare the rescue mission.

A scheme was devised to determine Skylab's roll attitude by using external temperature sensors located at 0 and 180 degrees on the cylindrical OWS. When the opposite sensors showed even heating, we were "wings level." If the temperature sensors differed significantly from one side to the other, it indicated that the OWS had rolled and the relative temperatures gave us a rough estimate of its orientation. The EGIL instructed the GNC flight controller to command the spacecraft's small roll jets to fire and re-balance the temperatures. This repetitive operation kept tension at the EGIL console at a high level. Who ever would have guessed that an EGIL systems guy would end up directing spacecraft attitude?

On top of all that, if the attitude of Skylab drifted too far, the sun-angle on the ATM's solar panels became too small to generate enough electricity to charge its batteries, causing any or all of its sixteen CBRMs to automatically trip off line. Radio contact with the spacecraft was not continuous; sometimes we had hour-long periods

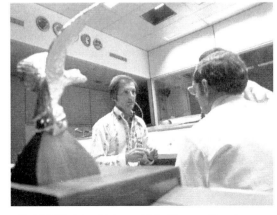

Skylab EGIL Sy explaining to Flight Director Glynn Lunney

Skylab EGIL Sy explaining to Flight Director Glynn Lunney with Rod Loe listening in

of no data and the bird would come into view with all of its CBRMs tripped off line, requiring the EGIL to send multiple commands to reconnect the power units. Sending commands using the push button system we had was very slow, and despite the urgency we had to take care not to make a mistake because with the spacecraft in such a precarious state any failure might result in its loss.

On one occasion, I was hurriedly commanding Skylab's electrical system and I skipped a step by not waiting for a Message Acceptance Pulse, which we called the MAP, to be received before I sent the next command. The consequence of this mistake was that one command got scrambled and transformed into something different. I soon found out what that was when the GNC flight controller reported to Flight Director Chuck Lewis that the bearing heaters in the large three-foot-diameter Control Moment Gyros that stabilized and turned the Station had been mysteriously turned off. It didn't take the INCO long to figure out that the spurious command had originated from my console (albeit scrambled). I was most embarrassed. Lewis ordered me to compose very precise commanding procedures and place them on all the MOCR consoles.

> It was like writing on the blackboard 100 times, "I shall not send bad commands to Skylab."

Because the Flight Control Team was required to monitor and control the Skylab 24/7, some "bright" person decided that a schedule known as "rotating shifts" would be best. This entailed working 8 a.m. to 4 p.m. for five days, then two days off; then 4 p.m. to midnight for five days, then 2 days off; finally midnight to 8 a.m., then 4 days off. We flight controllers maintained this rolling three-shift schedule throughout Skylab's orbital operations, and it slowly wore us down, both physically and mentally. Family life suffered, and in some cases, divorces resulted.

As the months passed, our tempers grew short and people became thin-skinned. Among the EGILs, "wars" were fought by way of critical personal entries in the ongoing console log. It would begin with a seemingly innocent comment or criticism, often no more than a three-or-four line entry concerning the actions of the EGIL on a previous shift. On his next shift, that EGIL might respond with as much as a four-*page* defense of his actions and make a criticism of the EGIL who made the original comment. Of course, this necessitated retaliation, and on and on it went. I advised one of the Flight Surgeons that psychological problems were developing among the ground personnel, to which he expressed genuine surprise; so dedicated was he to the medical and psychological needs of the space crews that he was clueless as to what was going on around him. I told him that he at least ought to investigate the condition that had developed on the ground, because we would eventually need to apply the "lessons learned" information to a future space station operation. It was a missed opportunity. During their 84-day mission, the final Skylab crew developed a "nose-out-of-joint" attitude and, thinking that we were withholding information from them, they in turn limited their communications with us. I have in my files a collection of console log war excerpts from the EGIL console. As a matter of fact, an amateurish poem appeared in the log, EGIL author unknown:

A Gathering of EGILS

The EGIL Team is a super team
Full of fire and full of steam
They quarrel, they fight, and curse each other.
And each thinks himself above the other.

First is Charlie, with his fatherly pose
Thinks himself senior, a primly pose
Write not ill of this stately one
For he takes offense and of you makes fun.

Next is Craig, essence of perfection,
He knows no flaws, his sense of direction
Leads him to see he's the truest knight
Ever to grace the space limelight.

And then there is Sy, noble and proud
His mind is dull, his mouth is loud

And when he sees the slightest open
He launches his barbs, and keeps on hopin'.

Least but not last is mighty Steve.
A word from him and one wants to leave.
He remains aloof, not mixing with others
He would really vanish, if I had my druthers.

Finally we come to righteous Moon
Who does no wrong, and thinks that soon
He will certainly rise above all others
And will take his place with the gods, his brothers.

Now you have seen how EGILs consist
And one wonders how they even exist
Together at all, with their petty selves,
So we hope tomorrow they will be on shelves.

The "sufferin'" Skylab EGILs were: Craig Staresinich, Steve McLendon, Sy Liebergot, Bill Moon, and Charlie Dumis.

I hated the Skylab Program.

A couple of the Skylab Flight Directors became involved in what I characterized as *Clash Of The Titans*. Don Puddy and Neil Hutchinson seemed to compete with each other. They created a blizzard of paper, and made last-minute changes to the minutely detailed daily flight plans that were sent up to the spacecraft by teleprinter, making our jobs more difficult.

Neil Hutchinson was a brilliant young man who seemed to possess a clear and precise vision of where he wanted to go and how he would get there. He started in the bowels of Building 30 as a Computer Supervisor in the RTCC. Computer Supes usually had little or nothing to say during simulation debriefings, but when I first noticed him during AS-201 Neil stood out by being very talkative, sometimes rivaling even the FIDO in holding the stage. I was surprised and a bit impressed when I saw him operating at the GUIDO console when AS-501 training began.

During the actual flight of AS-501, an unmanned test flight, Neil was to send a State Vector Update command to the CSM. The update was in the form of 16 lines of code to be sent all in one load. However, there was difficulty in sending this command load. Each time he failed to successfully uplink the update, Neil would announce this dramatically on the Flight Director's loop. As the tension mounted, Chris Kraft leaned over his console and said, "Keep trying, Guidance." And Neil would reply, "I am, Flight, I am. Looks like I'll have to send it line-by-line." This he did, dramatically announcing each successful code line transmission. When the short mission was over, I walked down to the Guidance console and observed to Hutchinson, "You sure got a lot of mileage out of that state vector update." He replied, with a sly smile, "Didn't I?"

He had gotten himself noticed again.

During Skylab, his seeming practice of always attempting to look good irritated some back room flight controllers, who created characterizations. Bill Greenshields, in the SSR would come up on the EGIL loop with, "whoosh … whoosh … whoosh." The EGIL would ask, "What's that?" Bill would reply, alluding to Neil, "Why, it's the Accelerated Man!"

As a Shuttle Flight Director, Neil would listen in on the systems flight controllers' intercom loops so that he could get a leg up on any problems and always look good. We knew he was doing it, and a scheme was concocted to sabotage his practice of eavesdropping. One day, during a shuttle orbit simulation, EECOM Bill Moon and his SSR EPS planned a bogus fuel cell problem to trap Neil. As the sim progressed, Moon called, "EPS, do you see that blip with that fuel cell?" As on Apollo, there were three fuel cells. On cue, the reply was, "Yep, I see it, let's watch it for awhile." Periodically, Bill and his EPS guy would talk to each other about "that" fuel cell, knowing that Hutchinson was probably eavesdropping. Finally, Neil could stand the suspense no longer, and called Moon, "EECOM, how're you looking there? How're the fuel cells?" Bill truthfully replied, "The fuel cells are absolutely nominal, Flight; no problems." *Gotcha!*

As his abilities were recognized, Neil continued to climb the success ladder. He retired from NASA, and is now a senior vice president for Science Applications International Corporation (SAIC) in San Diego. *Mission accomplished.*

Lots of good science came from the Skylab Program, especially in the field of solar astronomy, but for us flight controllers it was a year-long drudge. Ironically, it was solar activity that brought the orbiting laboratory down prematurely.

The final crew departed Skylab on February 8, 1974. After a couple of weeks of off-nominal systems engineering tests were conducted, it was placed it into a stable attitude and then its systems were shut off in the expectation that it would remain in orbit eight to ten years. However, the Sun was coincidentally at the minimum of its 11-year cycle of activity. As the activity increased, the Earth's atmosphere inflated and aerodynamic drag on Skylab increased, causing its orbit to decay more rapidly than expected, and so, in 1977, NASA began an effort to forestall the inevitable.

Fortunately, the solar panels were "hard wired" to the command receiver, so Skylab could be re-activated as long as there was solar power. On March 2, 1978, a small team of flight controllers went to the Bermuda tracking site and successfully re-powered the command receiver and took a "snapshot" look at the essential Skylab systems. They then returned to Houston to devise a plan to put Skylab in a minimum atmospheric drag attitude in order to extend its orbital lifetime as long as possible. The hope was that the Space Shuttle would be ready in time to reboost Skylab, both to prevent its uncontrolled entry and to save it for possible future use.

For nearly a year, the group of flight controllers worked 24/7 to refine and execute the plan, with little fanfare or recognition. A remotely-piloted reboost motor was cobbled together from off-the-shelf parts left over from Apollo, Jack Lousma and Fred Haise trained to fly it from the Shuttle, but it was to no avail. Development delays kept the Shuttle from saving the day, and on July 11, 1979, Skylab finally fell uncontrolled to Earth in a flaming re-entry that scattered debris over the Indian Ocean and a sparsely settled region of western Australia.

The unsung flight controller team consisted of Charlie Harlan, Bill Peters, Jim Saultz, Steve McLendon, Harry Clancy, Joe Fanelli, Debbie Dingle, Bill Gravitt, and Chuck Holliman.

Skylab simulations with EGIL Steve McClendon at the EGIL console

— Chapter 51 —
ASTP: The Russians are Coming...

The Apollo-Soyuz Test Project (ASTP) was the first international space flight. It was a symbol of the new spirit of "détente" between the two superpowers. The mission involved a linkup between an American Apollo CSM and a Russian Soyuz spacecraft and gave Deke Slayton and Vance Brand their first opportunity to fly. Tom Stafford was the commander. The Russians were so intent on a successful mission that they had a second launch vehicle complete with Soyuz spacecraft and crew, which was to be launched within 24 hours if the first launch failed. The Soyuz spacecraft was launched on July 15, 1975, followed by Apollo seven hours later. The docking occurred on July 17th, and after the hatches had been opened, the two commanders made an historic handshake for the TV camera. Joint operations were conducted for two full days, then the vehicles undocked and returned to Earth. In principal, this demonstrated a docking system that would enable the two nations to rescue one another's crews, but in fact, this would be the final Apollo mission.

The Soyuz was pressurized with a 14.7 psia normal Earth atmosphere, but we still used a 5 psia 100 percent oxygen environment on orbit, so the Docking Module / Airlock was built by us to allow the safe transfer of the crews between the spacecraft. After reaching orbit, the CSM retrieved the Docking Module from the top of the S-IVB and prepared for the docking with Soyuz.

Our first contact with the Russian space officials was on October 24, 1970. The Russians were a suspicious lot, so everything about the mission had to be negotiated, written down, and signed with multiple signatories. I was assigned as the Lead EECOM. I was delighted. The mission promised to provide exciting new experiences both technically and culturally, because I would have an opportunity for my first trip to Europe. On September 14, 1974, a group of us flight controllers endured a 31-hour trip for a two-week

ASTP: MOCR Identity Badge

ASTP: Soyuz spacecraft in flight before docking

ASTP CSM/Docking Module photo taken by the Soyuz cosmonauts

stay in Russia. Included in the group that finally assembled in Moscow were Steve Mclendon, Joe DeAtkine, Chuck Lewis, Chuck Deiterich, Jack Kamman, Bob Becker, Leo Reitan, Maurice Kennedy, Harry Black, Ed Tarkington (E.T.), astronauts Bo Bobko, Bob Overmyer, and a few others. Our visit was for training and familiarization. Later, a group of Russian flight controllers came to Houston for the same purpose. In order to cover any unforeseen contingencies with each other's spacecraft, a team of NASA flight controllers would be stationed in the Russian control center at Kaliningrad, near Moscow, and a team of Russian flight controllers in Houston Mission Control.

I was a bit apprehensive about my trip to the USSR, because of my Russian heritage and my ignorance about what the KGB might know of my family. My paternal grandparents emigrated from Kiev in the Ukraine in 1900, and my father, Solomon, was their first born in the US in 1910. My maternal grandparents emigrated from Novosibkov, Russia in 1913, together with my 4-year old mother, Ida. I consulted with our security personnel and was assured that there would be no problem.

We checked into the 4,000-room Hotel Rossia in Moscow and plopped exhausted into our small, but clean rooms. Bob Overmyer had previously warned me about the rooms being "bugged." He related that one day, on a previous trip, he opened the armoire looking for clean towels, but there were none, and grumbled aloud about this omission. Later that day, clean towels appeared. He said that it worked every time, i.e., "just yell into the armoire." I was to have a similar experience. Each floor had a "Key Lady" whose central location provided her with a commanding view of all hallways. Hotel protocol required that you give her your room key when leaving the hotel, and retrieve it from her upon returning. No sneaking women into your room, evidently.

The first cultural difference that I encountered was that beer (peevah) was served warm. There's nothing like a warm peevah after a hard day's work! However, I never managed to get a good meal there; the good restaurants were hidden from us. One of the cultural blunders that I made was while having a cup of tea (chai) and spooned into it what I thought was sugar from a bowl, since the other bowl contained some unappetizing, brownish rock-like hard crystals. I had to spit out the tea because I had spooned a couple of teaspoons of *salt* into it! The dark crystals in fact were unrefined sugar, and we were expected to place a 'cube' between our teeth and suck the tea through the sugar. This is called *vprikusku*, tea with a bite.

ASTP: Hotel Rossia in Moscow, USSR

I had no Russian language preparation before the trip, but while there, in spite of myself, I managed to learn to count to five, say "hello," "good-bye," "please," and "thank you." Actually, one of the really important words to learn was *pectaupaht*, which spelled restaurant when transliterated. Importantly, I should have learned to ask "Where am I?" in Russian, though.

Because the many coordination meetings required frequent travel between our countries, *per diem* arrangements were worked out where the Russians paid us 10 rubles per day, and when they came to the U.S. we paid them 24 dollars. What a deal! They'd head for the department stores here and load up with shoes, clothing, electronics, etc. In Russia, the 10 rubles went a long way – there was just nothing we wanted to buy. All the best goods were in the "dollar stores" that took only dollars and it was against Russian law to take rubles out of the country. The rubles were not even good souvenirs.

It was difficult to carry on a real conversation with the Russians. Most Russians I met and conversed with played everything "close to their chest," and offered little. On the other hand, we Americans talked about everything. It was very disconcerting, and after a while I found myself becoming reticent to volunteer small talk.

We were given VIP treatment, and visited all the usual VIP sites: the Bolshoi Theater to see *Swan Lake* ballet; The Russian Circus which seemed to smell of 10,000 horses; the Trinity Monastery in Zagorsk, which is the seat of the Russian Orthodox Church and a show place; and Red Square where we could see the Kremlin and St. Basil's Cathedral. Unfortunately, Lenin's Tomb was closed for repairs. So were all the churches, but in their case, *it was not for repair.*

ASTP: Conquest of Space obelisk base

ASTP: Conquest of Space obelisk rocket

The Gum Department Store, located off Red Square, served to demonstrate to us how poorly made the Russian consumer goods were; amazingly most of the store clerks used an abacus to count. We recognized that the Russian "spooks" were always near by; there ostensibly for our safety. Of course we had to visit the Exhibition Park For Economic Achievement, some 540 acres of permanent pavilions depicting Soviet "productivity."

On our way to the Exhibition Park, we encountered one of the most impressive monuments I have ever seen. It was the Conquest of Space obelisk; a rocket standing on a 200-foot tall swooping rocket exhaust plume made of titanium. The square base of the commemor-ative structure was 20 feet high and at least 100 feet on a side; each of the sides depicted the chronology of manned space flight with bas-relief figures of the engineers, cosmo-nauts, and the flight controllers.

We went to the Exhibition Park to tour the Space Pavilion and received a large dose of Russian space "firsts." None of us mentioned our rather significant "first" of landing men on the Moon, nor did our hosts bring up the subject. After a tour through the Gagarin Museum, it was on to lunch at The Corn Cob restaurant in the Park, where many vodka toasts were made.

Our Russian Mission Control host was Viktor Blagov, who at that time was a shift Flight Director. I swear he looked like Jack Palance. Two secretaries joined our official group, and attached themselves to the only two single guys – Steve McLendon and me ... hmm?

We were bussed outside of Moscow to the Russian Control Center, and entered through a large gate, replete with armed military guards. Our tour of the Control Center was revealing: while a showplace, it was certainly sparser than ours; in fact, one of their data displays had only 19 lines of data, updated once a minute. Some of our console displays had hundreds of data points, most updated once per *second*. It was hard to believe that the Russians were really ever in the race to put men on the Moon.

A daily blur of classroom training sessions in the Soyuz spacecraft followed. How tiny the two-man Soyuz capsule appeared. As I lay in the trainer, it was difficult to imagine how three cosmonauts could have fit, even without their space suits.

The best food was in the control center cafeteria, with its wide-open space and gleaming stainless steel. The first day we ate there, as Steve McLendon and I were moving down the food line, Steve pointed out that there were clusters of small flags of nations on

ASTP Shift Flight Director Viktor Blagov as our host in Moscow

each table, but not one displaying an American flag. I couldn't believe it, but he was correct; I echoed his complaint. The next day, there was an American flag in each table's flag cluster.

Bob Overmyer told me of another of his experiences with "eavesdropping." He was attempting to move a comfortable chair in the dayroom in the control center to better see the eternal hockey game on TV. As he lifted the chair, he almost broke a wire that ran down one leg and disappeared under the carpet. Our Russian colleagues averted their eyes, shaking their heads in embarrassment.

Joint mission simulations between the two control centers and crews could be interesting and challenging. Ego reared its ugly head during one of the joint mission simulations. During the debriefing after a training run, the Russian Mission Director Alexei Yeliseyev, a former cosmonaut, kept referring to each of the space crew, astronaut and cosmonaut, by their last names. It was Leonov this, Brand that, Stafford that, etc. After he had apparently referred to Tom Stafford by "Stafford" one too many times,

ASTP: The Russian secretaries

ASTP: Sy in tiny Soyuz Descent Module trainer in Star City

ASTP Russian Mission Director Alexei Yeliseyev and U.S. Lead Flight Director Pete Frank

Tom interrupted Yeliseyev and stated, "Excuse me, that's Brigadier General Thomas ... P ... Stafford!" slowly biting off each word for emphasis. There was a moment of stunned silence, after which a classy Alexei Yeliseyev picked up as if nothing out of the ordinary had occurred, referring from then on to Tom as General Stafford.

Nearly five years were spent carefully working out detailed agreements for this joint space venture; neither side trusted the other implicitly. Joint agreements were hashed out in great detail, and signed by authorities on both sides. During one of the final joint simulations, before docking the two spacecraft, to our surprise it was determined that the Soyuz was performing some unusual maneuvers. Flight Director Pete Frank inquired of the Russian Flight Director what they were doing. Pete was informed that the Soyuz was flying around the CSM / Docking Module assembly taking photographs. "We didn't know you planned to do that!" Pete protested. *"You didn't ask,"* was the simple Russian reply.

ASTP was to be the last Apollo flight, and there was a desire to prolong the mission of the CSM as long as safely possible. The joint mission was complete with the landing of the Soyuz after 6 days, but the CSM was to linger in orbit for another four days. The additional mission duration would stretch the cryogenic hydrogen for the fuel cells almost to fumes, so it would require close monitoring during the flight. Entry simulations allowed me to practice with this concern by depleting the hydrogen according to the computer model prediction. The SPS de-orbit burn was to occur on the other side of the Earth, out of telemetry contact, and in the event that it didn't happen, and a "Go Around" rev was required, the limited fuel cell hydrogen would necessitate an immediate power down by the crew in order to conserve the remaining gas. During each entry simulation, per the agreed-to procedure, I would pass an Equipment Emergency Power Down flight note to Lead Flight Director Pete Frank, the CapCom, The Trench guys, et al, just before the last communications pass ended. The note listed five major equipment groups, which were large power users and could be quickly turned off. We still would not "bother the crew" by having this simple procedure read up in advance or better still, have it already onboard in a checklist.

When the real mission last spacecraft communication pass came up, I was astonished to see Pete Frank turn over the Flight Director Console to Frank Littleton, who was functioning as only a

ASTP: Wide shot of MOCR

"Joint Flight Director" with Russian Mission Director Alexei Yeliseyev in the Russian Control Center. Littleton had *zero* real-time flight operations experience much less *any* console experience, yet there he stood at the Flight Director Console, uncertified. Well, before the CSM "went over the hill," I prepared my Equipment Emergency Power Down flight note to be passed up to the spacecraft, handing the first copy to Littleton. He looked at the note, and then blankly at me, asking, "What's this for?" I patiently explained to him the contents and purpose of the note and its importance to the safety of the crew in the event that the de-orbit burn did not come off and they had to go around. He replied, "We're not going around," disdainfully throwing the note on his console. Flabbergasted and infuriated, I clamped my jaw shut and sat down at my console. Inexplicably, Pete did not intervene. Fortunately, the reliable SPS engine performed flawlessly and the CSM de-orbited safely. To this day, I have not been able to fathom how Pete Frank was able to relinquish the Flight Director responsibility at that critical mission juncture, and then essentially walk away.

In my opinion, it was a singularly irresponsible act.

The flight crew almost perished during entry, once again proving the wisdom of New York Yankee baseball legend Yogi Berra's famous saying, "It ain't over till it's over." It has been reported that Vance Brand forgot to throw the two switches that activated the automatic functions of the Earth Landing System (ELS), an error that allowed deadly corrosive hydrazine fumes to be ingested into the cabin from the RCS jets that continued to fire and resulted in the failure to deploy the chutes. Tom was reading out the checklist that called for the actuation of the two ELS switches at 24,000 feet altitude. Vance stated publicly that he could have missed hearing Tom Stafford reading the steps in the checklist. However, he was monitoring the altimeter, and performed all the backup switch functions successfully. The three men suffered lung damage, and spent two weeks in a hospital while their wounds healed. Overall, the astronauts were to blame, because it was their concern that the chutes might inadvertently deploy in orbit, despite the fact there was no single point failure that would allow that to happen, that led to those switches not being enabled until the Command Module

ASTP: Sy holding the plaque with the two ELS switches that were not actuated on time after re-entry that he presented to Vance Brand at the Splashdown Party

descended to 24,000 feet.

It was a poor end to the final Apollo mission.

As a memento of his flight, I presented Vance Brand with a small walnut plaque to which were attached the two ELS switches that were overlooked during the ASTP re-entry. We had splashed down and the Russians had thumped down; the "locals" gathered at the JSC Gilruth Recreation Center for a post-flight cocktail party. I stood in a circle of a mixture of Americans and Russians making small talk. One of the Russian engineers asked me what we were going to do next. I replied, "We are going to weld the factory doors shut, disperse the engineering team, and start all over in five years with the Shuttle," which has since been characterized as an Apollo-era obituary. He declared us crazy, since they planned to launch six more times before year's end. I acknowledged the correctness of his observation, replying that it was simply how we did things in the U.S. The Russians pragmatically are still launching the same booster designs while we continue to pursue new, highly technical and increasingly expensive rocket designs.

Three of the huge Saturn V boosters remained unused after ASTP. Although built to boost humans into space, they instead each became a huge, expensive lawn ornament at JSC, MSFC and KSC.

Oh, how reliable and powerful the Saturn rockets were.

ASTP: Celebrating mission completion with vodka in the MOCR

Part Three

Life After Mission Control

— Chapter 52 —
EECOM No More: Set Adrift

In 1975, after more than a decade in the Mission Control Center, during which I had been involved in 11 lunar flights, Skylab, and ASTP, I began work on the new Shuttle spacecraft. I was assigned Lead EECOM duties to begin developing our Shuttle knowledge base, procedures, and mission rules. Once again I found myself at the beginning of another space program and spacecraft.

I was daunted by the task of learning the life support systems of a vehicle that was several times more complex than the Apollo CSM, and I must admit that I began this task with less enthusiasm than I'd had on the other programs. However, my interest picked up when I noted a capability deficiency in the water overboard dump system. The Orbiter had two overboard dump lines, one for waste water and the other for excess potable water generated by the fuel cells. Also, the first orbital flight would carry ten full 20-gallon tanks of water for equipment and crew cabin cooling, and I was concerned with the potential of a sizable water leak. I designed the Free Water Collection System that would permit water to be collected from anywhere in the cabin and vented directly overboard. In the event that one dump line became blocked, a cross-connection hose was included so that the other dump line could be used. The system found frequent use and is still part of the Orbiter systems. I felt some of the same satisfaction influencing Orbiter spacecraft system design as I did with the doubling of the critical crew switches on Apollo.

My enthusiasm waned however, as the pace picked up, review meetings became more frequent, flight documentation had to be created, and complex spacecraft systems operation had to be absorbed. In 1976, I reluctantly faced the realization that I no longer had the zeal for the real-time job that I had sustained for nearly a dozen years, and so I decided that it was time to "retire" as an active flight controller. I told Rod Loe, my branch chief, of my feelings and asked him to take me off the manning list for the first Shuttle test flight. He did just that. Unfortunately, there was no place in the Mission Operations Directorate for a person who no longer desired to be a flight controller, no matter how much experience he possessed. For the next three years, I was literally trapped within the organization and I drifted without a real job.

In 1979, salvation came in the form of the Space Station. Al Louvier gathered forty engineers together in Building 226, a small building in a back part of the JSC campus, where he informed us that we were to write a program plan for a space station. We were to start with a blank sheet of paper. I was excited; I'd finally be able to apply my hard-earned operations knowledge at the inception of a new program. Hal Loden, a former LM Control flight controller, was also assigned, and together we wrote the Operations section of the Space Station Program Plan.

A Space Station Program Office was created at JSC in 1981, and I became part of the Customer Integration Office under Carl Shelley, one of the most underrated experienced ops guys I ever knew. Life was good, every day tested my imagination and manned flight operations experience. I looked forward to going to work again.

Jim Beggs became the NASA Administrator and instituted a new "look" to NASA for the space station program in the form of the Work Package System. The space station design and development was divided into four "packages" and spread among four NASA centers: JSC in Houston, MSFC in Huntsville, KSC in Florida, and Lewis (since renamed Glenn) in Ohio. The Work Package concept was in marked contrast to the "Centers Of Excellence" approach to manned space flight, by which: JSC designed, developed and flew manned spacecraft; MSFC designed and developed the launch vehicles; and KSC performed the launches. The intent of the Work Packages was to "spread the wealth" by assigning more manned systems responsibility to MSFC, but the result was overlapping responsibilities, overspending, and angry Texas congressmen.

1986 found the space station program taken away from JSC and inexplicably moved to an expensive office complex located in Reston, Virginia, in the vicinity of Washington, D.C. Was a new NASA Center being created? The new space station, now named *Freedom*, commanded the attention of hundreds of personnel in Reston for eight years, consumed eight billion dollars, and yet produced no hardware. Andy Stofan, the new Space Station Program Director at Reston gathered together all 166 of us in the outgoing program office and told us in so many words, that he had no use for any of us. With my job eliminated, I was adrift again. Two years later, in 1988, I opted for early retirement, and so brought my career as a proud member of the NASA community to an end.

George Abbey became JSC Center Director in 1995 and the nexus of the space station moved back to JSC, where it has remained, though wounded by the wasteful eight-year stay in Reston.

Chapter 53
1972: Beginning Anew ... Several Times

After the divorce from Deanna on September 22, 1972, I moved into a small, unfurnished apartment at the Gaylord Terrace complex in Webster, decorated it in garish black, red and white colors, and installed a faux zebra-skin sofa and red shag carpeting that were (at least to me) masterpieces of interior decorating. Initially, I reveled in the solitude that my new single life afforded, but within a couple of months I discovered the other side: loneliness. When I socialized in the immediate area of JSC, I invariably ran into someone who told me they had seen my ex-wife accompanied by so and so, and I'm certain that she heard similar stories about me.

"We're going to the Space City Ski Club (SCSC) meeting tonight," Joanie McGinnis and Barbara Present, two lady friends told me. Why don't you go, too?" Ski club? I didn't snow ski, water ski, or any ski. But that night, I followed them to Sonny Look's Restaurant in Houston, just off the South Loop 610 Freeway.

I was to encounter the next big change in my life.

That evening in November 1972, I walked into a sea of people, my age or younger, talking excitedly to one another. All of these people snow skied? The vitality I experienced immediately buoyed me, but to my disappointment no one talked to me. My friends had disappeared into the crowd, so I made my way to the Membership table and optimistically joined the club. The next February, I went on my first ski trip to Steamboat Springs, Colorado. After two days of skiing lessons, I knew that I had found a new sport into which to pour my energy. Each year thereafter, I went on as many as five trips per winter, up to thirty days of skiing each year, and quickly became proficient.

When it became apparent that my new social life would be in Houston, I soon moved to Parkgate Apartments on Memorial Drive near downtown. In the large "singles" community of SCSC, my social life blossomed. I learned social graces, became a gourmet cook, oenophile, and resumed playing my classical guitar. Dinner at Sy's place was always a great evening.

My aggressiveness stood me in good stead as I soon volunteered to run a ski trip, was elected VP of Membership, and in my fourth year of membership, was elected President of the club. Finally, I became President of the statewide ski club organization called the Texas Ski Council.

You just couldn't keep a good flight controller down.

Six years after my divorce I met Tania Andrasko, a fine lady, and I married for the second time. Unfortunately, I once again found that I could not sustain a long-term relationship and after six years, on October 15, 1984, this marriage ended.

Chili Cook-offs are pure Texana. It all began in 1967 on the site of an old ghost town named Terlingua, on the edge of the Big Bend National Park in Texas. As a joke, Dallas newspaperman Frank X. Tolbert staged a cook-off between Texan Wick Fowler (creator of 3-Alarm Chili) and journalist / humorist H. Allen Smith. The results were a draw and the tradition was off and running, attracting more than 7,000 spectators to Terlingua each year.

Bitten by the Chili bug in 1979, I formed a team of friends, who included Tom Mercer, Glen Bishop, Bill Kirk, "Easy" Thayer and Ron Smith, to enter the

Buffalo Snort Chili cooking team logo

1982: Cooking at the International Chili Championship in Terlingua, Texas. Sy Liebergot and fellow competitor Richard Inman

new arena of competition as "Buffalo Snort Chili." With tents and canopy covers, a chili cook-off resembled a combination of a church bazaar and a camping ground. The people were honest and unsophisticated, and the prevailing humor, as reflected in the official chili newspaper, *The Goat Gap Gazette*, was delightfully "1950s-era high school." Each chili club was known as a Chili Pod and was headed by a president, with the title of The Great Pepper. I achieved that esteemed position in 1983. Also, it wasn't long before Buffalo Snort Chili added BBQ to the name and the acquisition of a large BBQ smoker pit. In 1986 this new consuming hobby came to an end, allowing me to move from a 1950s culture back to the present.

Soon after my second divorce in 1984, I decided to again take up serious

1982: Sy Liebergot cooking at the International Chili Championship in Terlingua, Texas

study of the classical guitar, which I had been considering doing for a few years. I now lived alone in a spacious townhouse located in southwest Houston. I lovingly removed my Condal from its case and placed it in its stand and prepared the music stand with the Carcassi study guide. All was in place to begin anew. Two weeks later, I received a call at work from my townhouse management company to tell me that my townhouse was burning. I rushed home from JSC to see the firefighters putting the finishing touches on extinguishing the fire. A policeman directing traffic denied me direct access, even though I informed him that it was *my* home on fire. I was unprepared for the hysteria that threatened to overcome me. I outwardly complied with the officer, but set off to find a back way to the smoldering structure. It turned out that the townhouse next door had

Townhouse fire rear view. Next door home was source of fire and completely gutted. My BBQ pit is seen in front of my unit

caught fire and was completely burned out, taking more than half of my home with it. The fire had entered the room that had held all my space mementos and my guitar. The brave firefighters had saved as much that they could, but the Condal was a total loss ... *so much for revisiting the classical guitar.* Ironically, if the guitar had remained in its case on the floor it probably would have survived. A year-long nightmare of restoration began.

Personal relationships became less frequent after the second failed marriage. Then I met a lady who was to cause me to fully mature as a person. In part due to her own personal issues, she had a hypersensitivity to my aggressive personality, of which I had absolutely no notion. She constantly criticized me for my tendency to interrupt others, a trait that I vociferously denied until I realized that she was correct; I always seemed to push myself into any situation. When our relationship was at the breaking point, inexplicably I believed that it was important to save it. That belief caused me to begin a series of visits to a local clinic specializing in psychotherapy. After several months of being told that I was perfectly normal, I was handed over to another psychologist who, after listening to me describe my problem with the lady, recommended that I purchase a book titled *Your Perfect Right*, and be ready to discuss it the following session. There was a chart in the book which described the main personality types: Passive, Aggressive, and Assertive. I always believed that I was perfect: Assertive. Not! Suddenly, there I was, described in the chart to a "Tee" as Aggressive! I was essentially a selfish bully. I was embarrassed by the realization that my friends had tolerated me for years.

Townhouse fire interior of my office and the ruined Condal classical guitar still in the stand and its case on the floor

At our next session, my therapist told me that aggression is permissible while you're achieving goals, and declared that since I had achieved my major goals, it was time to change. His actual words were, "Knock it off." In the final session, I related the lady's criticism that I constantly interrupted conversations, without being aware of my actions. As I spoke, I noticed that the therapist was moving his mouth. I inquired as to what he was doing; he replied that he was repeating every word that I spoke as closely as he could. The reason for my interrupting was to jump in as quickly as I could to put myself forward. If I concentrated totally on the speaker by repeating his speech to myself, I would break the poor behavior. It worked. I sometimes forgot what I was going to say, but this was a small price to pay. I was "fixed." Nevertheless, soon after, the lady and I went our separate ways.

I've always wondered why I persistently pursued that year-long psychotherapy.

One more milestone to "getting it right" appeared with the advent of my 50th birthday. Some people may disagree, but turning 50 years of age is especially difficult for a male. As February 15, 1986 approached, all kinds of doom-and-gloom thoughts invaded my conscience: Life is passing me by. There were now fewer years ahead of me than behind me and suddenly I no longer felt immortal. Has my life been fulfilled? Have I accomplished all that I had desired?

Should I buy a flashy new sports car?

Fortunately, that day came and went uneventfully 17 years ago, and I got on with my life.

1990: Marriage on the Golden Odyssey cruise ship. L-R: Jesuit priest Jerry Wade, Craig Liebergot, Sy Liebergot

— Chapter 54 —
Finally Getting It Right

I met Elizabeth Craig Tharpe in 1988, while snapping pictures at one of the monthly ski club meetings. I was soon informed that she preferred the name Craig. A dark-haired beauty of Irish descent (and temperament), she demonstrated an openness and quick wit that immediately attracted me. I asked her to join me for a drink in the hotel bar after the meeting. We hit it off. Our relationship burgeoned. In 1990, we embarked on a two-week Mediterranean Sea cruise on a 450-passenger Greek liner named *Golden Odyssey*, where on November 9, 1990, in the Sea of Lyon, our marriage was solemnized by Jerry Wade, a Jesuit priest, and witnessed by accompanying friends. The Greek captain later presented us with a card that stated our marriage occurred on "Friday, November 9, 1990, at 7:15 p.m., in the Gulf of Lyon, Lat. 41° 38.9'N, Long. 3° 41.5'E." Thus began my new journey with a partner with whom I remain to this day. Finally, at age 54, after two failed marriages, and eighteen years of emotional and social stumbling on my way to becoming a mature adult, I was qualified to deserve a woman as good as Craig Tharpe.

> *It is symbolic that another ship with which I had been intimately involved was also named Odyssey* – Apollo 13. *Each ship has represented a sea change in my life.*

From the start, married life with Craig was a smooth partnership of love of each other, singing, gourmet food, entertaining, and fine wine. We had wisely spent the previous two years learning about each other, living with each other, and working out most of the kinks of our relationship. I finally found myself actually "working" at the marriage; I learned the art of give and take.

On Craig's first visit to my townhouse in 1988, I cooked a sumptuous meal, and then led her to the wine racks to select a wine for dinner. She spied three bottles of 1970 Chateau Lafitte Rothschild and inquired when I planned to drink those special bottles. I replied, "If I marry again, one bottle will celebrate my engagement, the second my marriage, and the third my marriage's first anniversary." I believe Craig set her cap then and there to capture all three bottles of that fine wine, all of which we duly savored on those three special dates.

Early in 1988, I again experienced the need for live music in my life and I began the search for another fine Spanish-made guitar. Alas, it was not to be. In my haste to be on time for a scuba divemaster training session in a local lake, I slammed the rear car door on my right hand, trapping my index finger, and lost most of the fingernail. I could not return to the classical guitar, lacking an essential fingernail. So, at the ripe age of 52, after a 38-year hiatus, I decided to begin singing lessons once more. I discovered Kent Banbury, a voice teacher, who was New York trained and a veteran of Broadway live stage productions as a singer-actor. The development of my baritone voice was slow, but steady, and after weekly lessons over the years, I can today call myself a singer. I discovered that certain songs would evoke strong emotion in me. Kent, noting that, responded that singing was a window to the soul. Had I finally found a way to peel away my armor and overtly connect with my emotions? Was this the final stop on my musical journey? Jazz quartet drummer, classical guitar, beautiful guitar destroyed in fire, loss of an important fingernail, and finally singing. Craig is a singer too, having sung with a small group when she was in college. She and I have sung many duets, performing for friends.

All coincidence? Did I have a choice?

I became a proficient snow skier, and I was able to share a love of snow skiing with Craig. She was a wonderfully experienced scuba diver, and with no prompting from her, I began down the road to becoming a scuba diver, finally completing my training as a Scuba Divemaster in 1989. We were able to share many dive trips and thoroughly experience all that the new underwater world offered. The epitome of our dive trips came in 1994 on tiny Sipadan Island, some twenty-five miles off the northeast coast of Borneo. One hundred feet from the beach the twenty-foot depth drops off vertically to 2,800 feet; one picks the depth to which to dive, and the sea life varies according to depth. Sipadan Island was also a turtle sanctuary; hundreds of large sea turtles inhabited the surrounding waters and the females were frequently seen spending hours digging the large holes in the sand in which they bury their eggs.

Scuba divers Craig and Sy pose at 60 feet down

— Chapter 55 —
Travels and Food

After Apollo, I finally had the time and inclination to travel. The "travel bug" began with a trip to Russia in 1975 as part of the Apollo-Soyuz mission. Rather than fly straight home, I and three other flight controllers, Steve McLendon, Jack Kamman and Joe DeAtkine, followed this up with a car trip that began in Vienna, led across Europe, and ended in Madrid, Spain. In 1980, Tania Andrasko and I spent three-weeks in the Far East that included Hong Kong, the Peoples' Republic of China, Thailand, Singapore and Malaysia. 1983 found me with friends on a tour of England, followed by a train-boat trip across the English Channel that continued through France. Another trip to France in 1987 involved a 2,600-mile road trip throughout that country, Luxembourg, and Switzerland. France called again in 2002 and Craig and I spent two weeks in Paris and Bordeaux. We discovered that a three-hour lunch in a bistro could easily allow the consumption of two bottles of wine.

I was awestruck when I walked onto the floor of the huge amphitheater of the Asklepieion of Epidaurus, located in the Pelopennese, Greece's southern peninsula. Noted for its wonderful acoustics, the 2,400 year-old theater inspired me to sing *Bring Him Home*, the moving prayer from the musical production, *Les Miserables*, which was based on the novel by Victor Hugo. I was surprised and warmed by the applause that I received from other tourists scattered among the rows of towering stone seating.

It was a magic moment.

I have always disapproved of people that I've witnessed tearing or ripping at historic objects in order to go home with a souvenir. However, I found myself guilty of a like behavior when:

    I pocketed several pebbles from Moscow's Kremlin and felt like a thief.
    I pocketed several pebbles from the Great Wall of China and felt like a thief.

I had given up thieving after my childhood experiences, but I rationalized that those vast countries would never notice the loss.

Tourist behavior varies the world over, and a couple of instances stand clear in my memory that were annoying at the time but that I now find amusing:

Happens all the time: As I prepared to photograph a Bangkok temple spire from a special artistic perspective, the back of the head of a strange tourist suddenly appeared in my viewfinder at the moment that I pressed the shutter release.

Doesn't happen all the time: I settled into my seat to enjoy a variety show in

1990: Singing in the amphitheater in Epidaurus in Greece

Shanghai and I was shocked to realize that a person behind me had removed his shoes and now rested his stocking feet on my chair back, bracketing my ears.

During my explorations of France, I have never met a rude French citizen, despite opinions to the contrary by others who have traveled in France. However, I have seen "ugly American" tourists in action. Their main characteristic is that they expect everyone to speak English, and they speak loudly and slowly (in English) never saying please or thank you, even in English. Fortunately (for them), English has become more common in Paris, especially in the service industries.

In Shanghai, people spat a lot, presumably to rid themselves of bad spirits. I discovered why children wore "crotchless" underpants when I witnessed a father hold his small son at arms length facing the wall of a building.

In a Hong Kong open-air market I came upon 1000-Year-Old Eggs, a traditional Chinese delicacy. They were in wooden crates, buried in a black soil and one was dug out for me to sample. They are raw duck eggs that are preserved by fermenting them for a period of 100 days in a mixture of garden soil blended with charcoal ash, pine wood, salt, and strong black tea. (I was relieved to learn that, contrary to popular belief, horse urine is *not* used to aid the fermentation process.) I shelled the seemingly hard-boiled egg, and fearlessly bit into it exposing a vivid green yolk surrounded by a white that was blackish amber with various shades of yellow, blue, and green, with a flavor disagreeably cloying. As the old Latin adage says, *de gustibus non est disputandum* (there's no accounting for taste)!

Beijing restaurant guests were amused by the "Flying Shrimp" that I launched across the table when I squeezed my chopsticks on it much too hard.

Foreign restaurant menus were always a challenge. In a bistro in Saintes,

France, I spied an interesting sounding appetizer listed as *ansiette de bulots*, of which I hadn't a clue. I was still clueless when I noted the menu English translation was *whelks*. Bravely, without a laptop and Internet access for instant research, I went ahead with the order. They turned out to be large snails served cold on a bed of ice. Delicious. As I said, there's no accounting for taste!

In France, Americans always give themselves away when they request *beurre* (butter) at a restaurant; the French never serve it with a meal.

During another trip through France, several friends and I traveled the Rhone Valley from Dijon to Nice. In the midst of the Burgundy region, we stopped at a village restaurant for lunch. I always looked forward to French food with great expectation, and that day was to be no exception. After being seated, I noted on the nearby wall that there was an old sepia-toned photograph of a man with a magnificent handlebar mustache. Imagine my amazement when our waiter turned up sporting a large, fully waxed handlebar mustache, as did the server, a young man in his twenties. Grandfather, father, and son? Most likely, but we resisted the temptation to ask. For the meal, we all opted for the *degustation* and the wine selections were deferred to me, as I purportedly was the most knowledgeable. I was overwhelmed by a full-page, single-spaced listing of the available Burgundies. After a hopeless study, I essentially closed my eyes, and pointed at the middle of the page. I also selected Beaujolais for our appetizers by a similar "expert" method. Our dark-coated waiter poured the Beaujolais for me to taste, and the following dialogue ensued: "Sir, I said, "the Beaujolais is *chilled*." "Sir," he replied, "here in Burgundy, we *chill* the Beaujolais. In fact, we also slightly chill the burgundy." Completely humbled, I turned my attention to the food, which as expected, was wonderful.

Travel broadens one's horizons.

— Chapter 56 —
1988: Life After NASA

Soon after retiring from NASA in the spring of 1988, I was hired by Rockwell International, for whom now-retired former Flight Director Glynn Lunney had become Vice President of Houston Operations. It was a familiar crowd because Chris Kraft had an office up on Rockwell's executive row. My new job was in the Shuttle Program Office at JSC working for the Program Director, ex-Shuttle astronaut, Brewster Shaw.

Five years later, my job was eliminated. In the private sector, unless there is another position to accommodate you, you are laid off. Thankfully, I was given a six-month advance notice in which to find another position, but to no avail. I naively believed that my "silver bullet" was Glynn Lunney, with whom I had worked so closely during our years of mission control experiences, but I was sadly mistaken. As my notice came to a close with all my personal prospects exhausted, I placed several calls to Glynn's office; I was finally told by his secretary that "Mr. Lunney was aware of my calls." He never returned my calls.

After all those years, that hurt.

The next eight months were an ordeal. I made 500 job applications, with no responses. It slowly dawned on me that age discrimination was alive and kicking in the aerospace sector. It was the first time that I had been jobless since the age of ten. I met Mike Flory, a private educator, who spent his valuable time retraining me in the field of Environmental Engineering, a field that is fraught with more regulations than engineering. In the interim, he took me on as a consultant while I drew unemployment benefits. Damn, but I hated having to draw the unemployment checks; to me it was an admission of failure.

Later that year, Ed Fendell, retired ex-INCO flight controller and long-time friend, telephoned to inform me that Johnson Engineering Corporation (JEC), the company that operated the Weightless Environment Training Facility (WETF) at JSC, had been purchased by three people, two of whom were very familiar to me: ex-astronaut Gene Cernan and a former North American Aviation boss of mine, Tom Short. I immediately called Tom and he granted me an interview with a couple of his supervisors, Charlie Hoover and B.K. Miller, and bingo, I was hired.

I was soon to begin yet another career.

And what an interesting career it has turned out to be. I was responsible for directing the design and fabrication of International Space Station (ISS) element trainers for the new Neutral Buoyancy Laboratory (NBL), a water-filled pool that is used to simulate the conditions of weightlessness for astronaut training. The facility provides high-fidelity simulations of crew on-orbit operations in a controlled environment and substantially reduces their risks when performing EVA operations. In order to accommodate the large trainers, the NBL was made 200 feet long, 100 feet wide, and 40 feet deep, as compared to its smaller WETF predecessor, which was 78 feet long, 33 feet wide, and only 25 feet deep.

I derived tremendous satisfaction from seeing the full-scale trainers for the first ISS Node and its attached Pressurized Mating Adapters first designed as a computer generated form, then built, and finally used by the STS-88 astronauts who would mate the corresponding flight articles with a waiting Russian-built module. On December 4, 1998, I traveled to KSC and witnessed the launch of these elements, now named *Unity*.

After 34 years, it was the first and only launch that I had ever witnessed.

Tom McGowen, a retired naval officer and mechanical engineering professor, joined me the month after I was hired. Tom was far more experienced than me at building and designing things mechanical, but I complemented him with my spacecraft knowledge and familiarity with the workings of the Johnson Space Center, of which he knew little. He learned quickly. Aided by a shared whimsical sense of humor and a desire to get on with the new job, we became friends. Together we made a good team; we were Mutt and Jeff, Tom at six feet four inches and me at five feet eight inches, to the amusement of our co-workers.

Mutt and Jeff: Tom McGowen and Sy Liebergot

One day, during lunch, I accompanied Tom to a foreign car repair business where he was given the bad news by the young mechanic that the water pump on his seven-year-old Mercedes was in need of replacement. I was still wearing my JSC identification badge, that displayed both my photo and name. The young man kept staring at my badge as he talked to Tom, finally blurted out, "Are you THE Sy Liebergot? Did you know my father during Apollo?" Tom struggled to contain his laughter; and immediately nicknamed me TSL (THE Sy Liebergot). On a few occasions when that same scenario repeated, Tom could hardly control his mirth. As I said, Tom knew little of our space program, but he learned quickly.

Nicknames are given, not taken.

In 1998, JEC and its work force of 250 people was sold to a small company named SPACEHAB, and as the Space Station workload increased the combined staffs grew to 850 people. Tom Short stayed on for three years as a Senior Vice President to provide the necessary continuity.

Swiftly, five years passed; nearly all the NBL ISS EVA trainers were designed and built, and then the engineering work slacked off drastically. It is the vagary of aerospace engineering that when a project comes to an end, the people who delivered on the company's contract are either transferred, are laid off, or they resign. I knew my days were numbered, and since my relationship with Tom Short was on a personal basis, I prevailed upon him to help me find another position. Once again he came to my rescue, and found a position for me in the ISS Configuration Management organization of the company. As a Senior Configuration Management Analyst, I keep track of the equipment that is moved on and off the orbiting International Space Station, a job that challenges all my experience and knowledge of spacecraft hardware and space operations.

Once again, another new career.

For most of my aerospace career, I was in flight operations and served as a flight controller, and in that capacity I competed with the Mission Evaluation Room (MER) engineering community for expert knowledge of the spacecraft systems. Such was the arrogance of a flight controller that he set out to know as much or more about the systems for which he was responsible than the people assigned to support him. Ironically, my new job places me fully on the other side of the fence as an engineer in the MER where I find myself now criticizing the Flight Control Team for their arrogance and lack of cooperation.

Late in life, I have discovered that every fence has two sides.

Epilogue: *Reflections*

When Neil Armstrong and Buzz Aldrin first stepped out on the soil of an alien planet, I thought, *"Mankind was exploring again."* The United States had stepped forward to take its place as a great exploring nation, replacing those nations which, having lost the will to be great, had fallen aside.

In the European, Japanese, and Russian participation in the International Space Station, I see glimmerings of others readying to join the small space exploring community and the Chinese have demonstrated strong ambitions to put men in space, and ultimately on the Moon.

However, the promises for the future that 1969 held have not borne fruit. It is sad to reflect that, as of this writing, more than 33 years have passed since that historic heroic feat on July 20, 1969 that highlighted the triumphant days of the U.S. space program, and we are not again reaching beyond our planet with manned spacecraft. Mars awaits, as does a base on the Moon. Life is replete with missed opportunities. Has our nation missed its manifest destiny in space?

I have never been certain of the direction in which I wanted my life to go. As I reflect on these pages, I see myself having traveled a road on which I constantly had difficulty staying, but there always seemed to be a nudge here, a bump there that kept me moving toward my own manifest destiny. That destiny is this journey that has allowed me to partake of the rich tapestry of life, a tapestry that is woven from the many acquaintances and friends I have met along the way who added to my life, dispensed invaluable advice, or set examples for me to follow. It is a unique weaving containing threads of incredible food, wine, sports, music and travel, most of which has been shared or inspired by my wife, Craig.

Thank you, Craig.

— **Appendices** —

— Appendix A —
The Mission Control Patch

The following words are attributed to Gene Kranz and were formalized by Pete Frank:

This emblem was developed for the Mission Control team to recognize their unique contribution to the Manned Space Program. To date, we have flown 47 missions during the Mercury, Gemini and Apollo Programs. These missions have succeeded due to the efforts of the Mission Control Team.

In selecting the theme of the emblem, we centered upon the Sigma as the dominant element. This sigma was used once before, for the *Mercury 8* spacecraft. The sigma represents the total mission team. In addition, it represents the individual flight control teams from all programs past, present and future. Within the teams, it represents all engineering, scientific, and operations disciplines and tasks in support of the spacecraft and aircraft program elements. To a great extent, it can represent many other things in consideration of the benefits for all mankind that are possible through space.

The rocket launch represents the dynamic elements of space, the initial escape from our environment and the thrust to explore the universe. The energy of the program must be maintained by the mission team if the space goals are to be achieved.

The remaining elements are the Earth, planets and the stars. The Earth is our home and will forever be serviced by both manned and unmanned spacecraft in order to improve the quality of life of our present home. The stars and planets represent a major source of scientific study as well as the challenge of exploration for the future mission control teams. At no time should we lower our sights below the quest for the stars, for only in this way are we chal-

Mission Control Emblem

lenged sufficiently to be better than we are.

The border of the patch contains symbols to represent the three major programs that have been supported by the team. The Mercury, Gemini and Apollo Programs have seen a succession of many great moments. These programs succeeded due to the dedication of the many people who formed the teams and committed their being to the team. The four stars represent the current and future programs: Skylab, Apollo-Soyuz Test program (ASTP), Earth Resources Aircraft Program, and the Shuttle. At the conclusion of these programs, modifications to the emblem will be made to replace the star with an appropriate symbol for the program.

The wording on the patch was chosen to stress the very positive attitude used by the Mission Control team to assure crew safety and mission success. "Achievement through Excellence" is the standard for our work. It represents an individual's commitment to a belief, to craftsmanship and perseverance. With the above qualities, a positive approach is created that assures objective accomplishment and the return of the crew.

The emblem thus recognizes your contribution to history, and a commitment to current and future programs, and will assure continuity of the great teams of the past.

Eugene F. Kranz — 1973

— Appendix B —
Apollo 13: O_2 Tank Chronology

An 8-year chain of events was required to make a bomb. This chain could have been broken by any one of these events averting the Apollo 13 *disaster.*

1962: North American Awards subcontract to Beech Aircraft to design and build cryogenic storage tanks. Specifications call for tank heaters and fans to use 28 VDC.

1965: KSC launch facilities use 65 VDC ground power. North American notifies subcontractors, but thermostat subcontractor doesn't get the word.

1966: Cryo O_2 tank manufacture begins. Includes 28 VDC heater thermostat

March 11, 1968: Cryo O_2 shelf (O_2 Tanks 1 & 2) assembly begins. It is slated for the *Apollo 10* mission.

June 4, 1968: Oxygen tank shelf installed in SM106, the *Apollo 10* Service Module.

October 21, 1968: When the O_2 tank shelf was being removed by hoist from the *Apollo 10* SM for modifications, a bolt snapped and it dropped about 2 inches. This impact probably misaligned an already loose fill / drain tube fitting.

November 22, 1968: Subsequent testing revealed no discrepancies, although no testing with cryogenics was conducted. The required modifications necessitat-

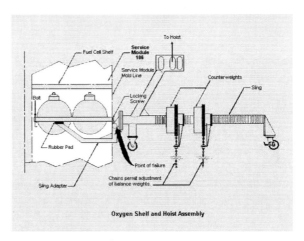

Apollo 13: hoist assembly that dropped the Apollo 10 cryogenic oxygen tank set

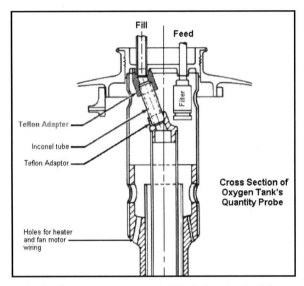

Apollo 13 cryogenic oxygen tank fill/drain line showing Teflon adapter that became cocked when the tank set was dropped

Apollo 13 SM Equipment Bay showing Cryogenic Oxygen tank shelf. O2 tank 2 is shown (in front)

ed reassigning the O_2 tanks shelf to a later mission, *Apollo 13*. One mod was not accomplished: changing the 28 VDC tank heater thermostats to ones with a higher rating to be compatible with KSC ground power of 65 volts.

January, 1969: The oxygen tank set is installed in the *Apollo 13* SM and completes and passes all testing.

June, 1969: The *Apollo 13* SM is shipped to KSC for further testing, mating to the launch vehicle and launch. At KSC the CM and the SM were mated, checked, assembled on the Saturn V launch vehicle and the total launch vehicle moved to the launch pad.

March 16, 1970: A major prelaunch pad test, the Countdown Demonstration Test (CDDT) begins. This is an "all up" test, which included filling the SM H_2 and O_2 cryogenic tanks. Difficulty with normal detanking (with gaseous oxygen) was experienced with O_2 Tk 2; while O_2 Tk 1 dropped quantity to 50 %, O_2 Tk 2 dropped to only 92%. This was probably due to misalignment of the fill / drain tube.

March 27, 1970: Detanking operations were resumed with a decision to try to boil off the remaining oxygen in O_2 Tank 2 by use of the tank heaters and fans, applying the 65 volt DC GSE power. The tank was emptied after 8 hours of tank heaters and fan operation, the effects of which were disastrous.

The heater thermostats, rated for 28 VDC and set to open at a tank temperature of 80°F, were fused closed when the KSC 65 VDC ground power was applied.

Anecdotally, a technician who was monitoring the boil-off tank drain procedure was told not to allow the tank temperature to rise above 80°F. However, the temperature gauge on his GSE panel could only read to 80°F, allowing continuous tank heaters operation for eight hours. During this time the internal tank temp rose to as much as 1,000°F, causing the Teflon wiring insulation to crack open, *exposing bare wires*.

March 30, 1970: It was decided to test the oxygen tanks again by filling them. Tests were satisfactory but Tank 2 still did not empty normally like Tank 1. The boil-off procedure was repeated and the heaters were again left on for hours.

Eight years after contract award, the stage was now set for our worst space disaster.

Apollo 13 SM Equipment Bay showing fuel cells 3 (left) and 1 on the top shelf. Oxygen Tank 2 can just be seen below

Apollo 13 cryogenic tank heater welded thermostat

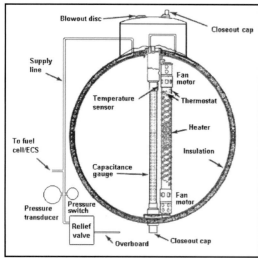

Apollo 13 Internal parts of the cryogenic oxygen tank showing fans, heater and thermostat

April 11, 1970: Launch of the *Apollo 13* mission.

April 12, 1970: Cryo tanks were stirred about 47 hours into the mission. O_2 Tank 2 quantity reading fails off scale at the high end. Cryo tanks stirred again about two hours later with no change in O_2 Tank 2 quantity reading

April 13, 1970: EECOM Flight Controller Sy Liebergot requested an extra cryo tanks stir at crew pre-sleep time to obtain a good reading on O_2 Tank 1 quantity. Stir commenced at 55:53:20.

April 13, 1970: O_2 Tank 2 blew up (over pressurized) 33 seconds later at 55:54:53, (9:06 p.m. CST).
So within seconds, this was the result:
o O_2 Tank 2 exploded, venting. SM panel blown off by 60,000 psi (7 pounds of TNT).
o O_2 Tank 1 leaking rapidly with 3 hours to empty.
o Fuel Cells 1 & 3 dead. Fuel cell 2 life is tied to O_2 Tank 1.
o Spacecraft gyrating wildly.

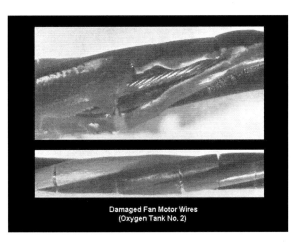

An example of the Apollo 13 cryogenic oxygen tank Teflon fan wiring cracked and bared after being subjected to excessive heater operation during the pad test

— Appendix C —

Before O$_2$ Tank 2 Exploded — Entry Battery B Charging

```
LM1885                    CSM EPS HIGH DENSITY                      0518
CTE  055:46:51 (        ) GET  55:46:53         (          )  SITE
─────── DC VOLTS ───────  ─────── AC VOLT ──────               ── FC  °F ──
CC0206    VMA     29.5    CC0200  AC 1    115.6      SC2084  1  SKN    409.1
CC0207    VMB     29.4    CC0203  AC 2    115.7      SC2085  2  SKN    412.7
CC0210    VBA     36.4    ─────── FC PSIA ──────     SC2086  3  SKN    414.5
CC0211    VBB     39.5 *  SC2060  1  N2    55.8      SC2081  1  TCE    158.0
CC0232    VBR     35.8    SC2061  2  N2    53.9      SC2082  2  TCE    158.9
CD0200    VMLA    0.15    SC2062  3  N2    54.4      SC2083  3  TCE    157.1
CD0201    VMLB    0.15    SC2066  1  O2    64.6      ────── FC  RAD  °F ──
CD0005    VMQA    0.15    SC2067  2  O2    62.7      SC2087  1  OUT     70
CD0006    VMQB    0.15    SC2068  3  O2    63.5      SC2088  2  OUT     71
─────── DC AMPS ───────   SC2069  1  H2    64.7      SC2089  3  OUT     75
          TOT  SC  67.7   SC2070  2  H2    62.9      SC2090  1  IN      86
          TOT  FC  67.6   SC2071  3  H2    63.4      SC2091  2  IN      88
       FC PCT SC 100.0    1  O2-N2    ΔP    8.8      SC2092  3  IN      95
          TOT BAT   0.0   2  O2-N2    ΔP    8.8     ─ PCT  TOTAL  FC  LOAD
       BAT PCT SC        3  O2-N2    ΔP    9.1              FC  1     31.6
SC2113    FC   1   21.4   1  H2-N2    ΔP    8.9              FC  2     31.6
SC2114    FC   2   21.3   2  H2-N2    ΔP    9.0              FC  3     36.9
SC2115    FC   3   24.9   3  H2-N2    ΔP    9.1      ──────── INST ────────
CC0222    BAT  A    0.0   ────── FC  LB / HR ─────   CT0120  PCM       HBR
CC0223    BAT  B    0.0   SC2139  1  H2   .0659      CT0125  4.25    4.249
CC0224    BAT  C    0.0   SC2140  2  H2   .0679      CT0126  0.75     .731
CC0215    CHRGR   1.12 *  SC2141  3  H2   .0739      CT0340  TMG       CTE
CC2962    LM       1.6    SC2142  1  O2   0.488      CT0015  +20      20.1
SC2160    PH  1   LOW     SC2143  2  O2   0.507      CT0016  -20     -20.0
SC2161    PH  2   LOW     SC2144  3  O2   0.550      CT0017   +5      5.03
SC2162    PH  3   LOW                                CT0018  +10      10.1
                                  1 ──── 2 ──── 3    CT0620   SS     -74.7
     CC0175/76/77   INV  TMPS    90    88    73      CS0220  PROBE   312 *
```

Entry Battery B Charging has been terminated

```
LM1885                    CSM EPS HIGH DENSITY                      0518
CTE  055:54:52 ( 55.914 ) GET  55:54:54         ( 55.915  )  SITE
─────── DC VOLTS ───────  ─────── AC VOLT ──────               ── FC  °F ──
CC0206    VMA     29.5    CC0200  AC 1    115.6      SC2084  1  SKN    409.1
CC0207    VMB     29.19   CC0203  AC 2    116.3      SC2085  2  SKN    412.7
CC0210    VBA     36.71   ─────── FC PSIA ──────     SC2086  3  SKN    412.5
CC0211    VBB     37.56   SC2060  1  N2    55.80     SC2081  1  TCE    157.2
CC0232    VBR     37.00   SC2061  2  N2    53.92     SC2082  2  TCE    158.9
CD0200    VMLA    0.15    SC2062  3  N2    54.08     SC2083  3  TCE    157.5
CD0201    VMLB    0.15    SC2066  1  O2    64.58     ────── FC  RAD  °F ──
CD0005    VMQA    0.15    SC2067  2  O2    62.69     SC2087  1  OUT     70
CD0006    VMQB    0.15    SC2068  3  O2    63.22     SC2088  2  OUT     71
─────── DC AMPS ───────   SC2069  1  H2    64.68     SC2089  3  OUT     76
          TOT  SC  71.6   SC2070  2  H2    62.60     SC2090  1  IN      86
          TOT  FC  71.6   SC2071  3  H2    63.44     SC2091  2  IN      88
       FC PCT SC  99.96   1  O2-N2    ΔP    8.78     SC2092  3  IN      93
          TOT BAT   0.0   2  O2-N2    ΔP    8.77        PCT  TOTAL  FC  LOAD
       BAT PCT SC   0.4   3  O2-N2    ΔP    9.14             FC  1     30.9
SC2113    FC   1   22.1   1  H2-N2    ΔP    8.88             FC  2     32.0
SC2114    FC   2   22.9   2  H2-N2    ΔP    8.68             FC  3     37.0
SC2115    FC   3   26.5   3  H2-N2    ΔP    9.36     ──────── INST ────────
CC0222    BAT  A    0.0   ────── FC  LB / HR ─────   CT0120  PCM       HBR
CC0223    BAT  B    0.0   SC2139  1  H2   .0671      CT0125  4.25    4.249
CC0224    BAT  C    0.0   SC2140  2  H2   .0685      CT0126  0.75    .7312
CC0215    CHRGR  -0.008   SC2141  3  H2   .0760      CT0340  TMG       CTE
CC2962    LM       2.3    SC2142  1  O2   0.402      CT0015  +20      20.11
SC2160    PH  1   LOW     SC2143  2  O2   0.401      CT0016  -20     -20.03
SC2161    PH  2   LOW     SC2144  3  O2   0.476      CT0017   +5      5.031
SC2162    PH  3   LOW                                CT0018  +10      10.06
                                  1 ──── 2 ──── 3    CT0620   SS     -74.7
     CC0175/76/77   INV  TMPS    90    87    73      CS0220  PROBE   312 *
```

Apollo 13: EECOM Console Displays

Before O₂ Tank 2 Exploded

```
LM1839                   CSM ECS-CRYO TAB                        0613
CTE 055:46:51 (      )   GET 055:46:53  (        )        SITE
———— LIFE SUPPORT ————                  ——— PRIMARY COOLANT ———
GF3571   LM CABIN P   PSIA              CF0019   ACCUM QTY   PCT     34.4
CF0001   CABIN P      PSIA      5.1     CF0016   PUMP P      PSID    45.0
CF0012   SUIT P       PSIA      4.3     SF0260   RAD IN      T   °F  73.8
CF0003   SUIT ΔP      IN H2O   -1.68
CF0015   COMP ΔP      P PSID    0.30
CF0006   SURGE P      P PSIA   891      CF0020   RAD OUT T       °F   35
         SURGE QTY    LB        3.67    CF1081   EVAP IN T       °F   45.7
  02     TK 1 CAP ΔP  PSID     21       CF0017   STEAM T         °F   64.9
  02     TK 2 CAP ΔP  PSID     17       CF0034   STEAM P       PSIA  .161
                                        CF0018   EVAP OUT T      °F   44.2
CF0036   02 MAN P     PSIA    105
CF0035   02 FLOW      LB/HR     0.181
                                        SF0266   RAD VLV 1 / 2        ONE
CF0008   SUIT T              °F  50.5   CF0157   GLY FLO     LB / HR  215
CF0002   CABIN T             °F  65     ——— SECONDARY COOLANT ———
CF0005   C02 PP       MMHG    1.5       CF0072   ACCUM QTY   PCT     36.8
———————— H20 ————————                   CF0070   PUMP P      PSID     9.3
CF0009   WASTE        PCT    24.4       SF0262   RAD IN      T   °F  76.5
         WASTE        LB     13.7       SF0263   RAD OUT     T   °F  44.6
CF0010   POTABLE      PCT   104.5       CF0073   STEAM P       PSIA .2460
         POTABLE      LB    37.6        CF0071   EVAP OUT T      °F  66.1
CF0460   URINE NOZ T         °F  70     CF0120   H20-RES       PSIA  25.8
CF0461   H20 NOZ T           °F  72     TOTAL    FC CUR        AMPS
———————— CRYO SUPPLY ———————   — 02-1 —   — 02-2 —   — H2-1 —   — H2-2 —
SC0037-38-39-40  P        PSIA   913       908      225.7 *    235.1
SC0032-33-30-31  QTY      PCT    77.63     01.17     73.24      74.03
SC0041-42-43-44  T        °F    -189      -192      -417       -416
                 QTY      LBS    251.1     260.0     20.61      20.83
```

O₂ Tank 2 Pressure 996 psia (Fire in the Tank)

```
LM1839                   CSM ECS-CRYO TAB                        0613
CTE 055:54:52 ( 55.914 ) GET 055:54:54  ( 55.915 )        SITE
———— LIFE SUPPORT ————                  ——— PRIMARY COOLANT ———
GF3571   LM CABIN P   PSIA              CF0019   ACCUM QTY   PCT     34.4
CF0001   CABIN P      PSIA      5.1     CF0016   PUMP P      PSID    46.0
CF0012   SUIT P       PSIA      4.3     SF0260   RAD IN      T   °F  73.8
CF0003   SUIT ΔP      IN H2O   -1.72
CF0015   COMP ΔP      P PSID    0.30
CF0006   SURGE P      P PSIA   841      CF0020   RAD OUT T       °F   35
         SURGE QTY    LB        3.67    CF1081   EVAP IN T       °F   46.2
  02     TK 1 CAP ΔP  PSID    -12       CF0017   STEAM T         °F   64.9
  02     TK 2 CAP ΔP  PSID    105 *     CF0034   STEAM P       PSIA  .161
                                        CF0018   EVAP OUT T      °F   44.2
CF0036   02 MAN P     PSIA    103
CF0035   02 FLOW      LB/HR     0.181
                                        SF0266   RAD VLV 1 / 2        ONE
CF0008   SUIT T              °F  50.5   CF0157   GLY FLO     LB / HR  214.6
CF0002   CABIN T             °F  65     ——— SECONDARY COOLANT ———
CF0005   C02 PP       MMHG    1.5       CF0072   ACCUM QTY   PCT     37.5
———————— H20 ————————                   CF0070   PUMP P      PSID     9.3
CF0009   WASTE        PCT    24.8       SF0262   RAD IN      T   °F  76.5
         WASTE        LB     13.9       SF0263   RAD OUT     T   °F  46.2
CF0010   POTABLE      PCT   104.1       CF0073   STEAM P       PSIA .2460
         POTABLE      LB    37.5        CF0071   EVAP OUT T      °F  66.3
CF0460   URINE NOZ T         °F  72     CF0120   H20-RES       PSIA  25.8
CF0461   H20 NOZ T           °F  72     TOTAL    FC CUR        AMPS  71.58
———————— CRYO SUPPLY ———————   — 02-1 —   — 02-2 —   — H2-1 —   — H2-2 —
SC0037-38-39-40  P        PSIA   879 *     996 *    224.2 *    233.6 *
SC0032-33-30-31  QTY      PCT    76.83     47.04 *   73.24      74.03
SC0041-42-43-44  T        °F    -190      -329 *   -417       -416
                 QTY      LBS    248.5     260.0     20.61      20.83
```

O₂ Tank 2 has Exploded
Fuel Cells 1 & 3 O₂ Flows dropping

```
LM1885                        CSM EPS HIGH DENSITY                          0518
CTE  055:54:56  ( 55.916 )   GET 55:54:58      ( 55.916 )        SITE
─────── DC VOLTS ───────     ────── AC VOLT ──────            ─────── FC    °F ──────
CC0206    VMA      28.65      CC0200   AC 1      116.2         SC2084  1  SKN    409.1
CC0207    VMB      28.48      CC0203   AC 2      116.3         SC2085  2  SKN    412.7
CC0210    VBA      36.71     ────── FC  PSIA ──────            SC2086  3  SKN    412.5
CC0211    VBB      37.56      SC2060  1  N2      0.17 *        SC2081  1  TCE    157.6
CC0232    VBR      36.82      SC2061  2  N2     53.92          SC2082  2  TCE    158.9
CD0200    VMLA      0.15      SC2062  3  N2     54.38          SC2083  3  TCE    156.7
CD0201    VMLB      0.15      SC2066  1  O2     64.28         ─────── FC   RAD  °F ──────
CD0005    VMQA      0.15      SC2067  2  O2     62.69          SC2087  1  OUT      71
CD0006    VMQB      0.15      SC2068  3  O2     63.52          SC2088  2  OUT      71
─────── DC AMPS ───────       SC2069  1  H2     64.68          SC2089  3  OUT      76
          TOT   SC   81.5     SC2070  2  H2     62.60          SC2090  1  IN       86
          TOT   FC   81.5     SC2071  3  H2     63.44          SC2091  2  IN       92
    FC PCT SC    99.97        1  O2-N2   ΔP     64.11 *        SC2092  3  IN       93
          TOT  BAT   0.0      2  O2-N2   ΔP      8.77           PCT  TOTAL  FC  LOAD
      BAT PCT  SC    0.3      3  O2-N2   ΔP      9.14                FC 1      31.6
SC2113    FC 1       25.7     1  H2-N2   ΔP     64.50 *              FC 2      32.0
SC2114    FC 2       26.1     2  H2-N2   ΔP      8.68                FC 3      36.4
SC2115    FC 3       29.7     3  H2-N2   ΔP      9.06         ─────── INST ───────
CC0222    BAT A      0.0     ────── FC LB / HR ──────         CT0120  PCM      HBR
CC0223    BAT B      0.0      SC2139  1  H2     .0775          CT0125  4.25    4.24
CC0224    BAT C      0.0      SC2140  2  H2     .0786          CT0126  0.75    .731
CC0215    CHRGR    -0.008     SC2141  3  H2     .0882          CT0340  TMG     CTE
CC2962    LM         2.5      SC2142  1  O2     0.193 *        CT0015  +20     19.5
SC2160    PH 1       LOW      SC2143  2  O2     0.421          CT0016  -20    -20.0
SC2161    PH 2       LOW      SC2144  3  O2     0.275          CT0017   +5     5.03
SC2162    PH 3       LOW                                       CT0018  +10    10.00
                                    1 ──── 2 ──── 3            CT0620   SS     -104
     CC0175/76/77    INV   TMPS    90     87     73            CS0220  PROBE   312 *
```

Fuel Cells 1 & 3 Dead (currents and O₂ and H₂ flows zero)
Main DC Bus B and AC Bus 2 dead

```
LM1885                        CSM EPS HIGH DENSITY                          0518
CTE  055:58:24  ( 55.973 )   GET 55:58:26      ( 55.974 )        SITE
─────── DC VOLTS ───────     ────── AC VOLT ──────            ─────── FC    °F ──────
CC0206    VMA      25.27 *    CC0200   AC 1      116.2         SC2084  1  SKN    409.1
CC0207    VMB       4.62 *    CC0203   AC 2        0.1 *       SC2085  2  SKN    412.7
CC0210    VBA      36.71     ────── FC  PSIA ──────            SC2086  3  SKN    412.5
CC0211    VBB      37.38      SC2060  1  N2      0.17 *        SC2081  1  TCE    158.4
CC0232    VBR      36.65      SC2061  2  N2     53.92          SC2082  2  TCE    159.7
CD0200    VMLA      0.15      SC2062  3  N2     53.49          SC2083  3  TCE    157.9
CD0201    VMLB      0.15      SC2066  1  O2     55.67 *       ─────── FC   RAD  °F ──────
CD0005    VMQA      0.15      SC2067  2  O2     61.20          SC2087  1  OUT      73
CD0006    VMQB      0.15      SC2068  3  O2     54.03 *        SC2088  2  OUT      77
─────── DC AMPS ───────       SC2069  1  H2     63.78          SC2089  3  OUT      81
          TOT   SC   58.2     SC2070  2  H2     62.30          SC2090  1  IN       87
          TOT   FC   58.1     SC2071  3  H2     61.96          SC2091  2  IN       91
    FC PCT SC    99.95        1  O2-N2   ΔP     55.49 *        SC2092  3  IN       96
          TOT  BAT   0.0      2  O2-N2   ΔP      7.28 *       ─ PCT  TOTAL  FC  LOAD
      BAT PCT  SC    0.5      3  O2-N2   ΔP      0.55 *              FC 1       1.4 *
SC2113    FC 1       0.8 *    1  H2-N2   ΔP     63.61 *              FC 2      91.8 *
SC2114    FC 2      53.4 *    2  H2-N2   ΔP      8.38                FC 3       6.9 *
SC2115    FC 3       4.0 *    3  H2-N2   ΔP      8.47         ─────── INST ───────
CC0222    BAT A      0.0     ────── FC LB / HR ──────         CT0120  PCM      HBR
CC0223    BAT B      0.0      SC2139  1  H2     .0068 *        CT0125  4.25    4.26
CC0224    BAT C      0.0      SC2140  2  H2     .1546 *        CT0126  0.75    .7510
CC0215    CHRGR    -0.008     SC2141  3  H2     .0089 *        CT0340  TMG     CTE
CC2962    LM         1.6      SC2142  1  O2     0.081 *        CT0015  +20     20.0
SC2160    PH 1       LOW      SC2143  2  O2     0.933 *        CT0016  -20    -20.0
SC2161    PH 2       LOW      SC2144  3  O2     0.075 *        CT0017   +5     5.03
SC2162    PH 3       LOW                                       CT0018  +10    10.00
                                    1 ──── 2 ──── 3            CT0620   SS      -94
     CC0175/76/77    INV   TMPS    90     85     73            CS0220  PROBE   312 *
```

O_2 Tank 2 has exploded (19)
O_2 Tank 1 leaking and has dropped to 782 psia

```
LM1839                        CSM ECS-CRYO TAB                          0613
CTE 055:54:56 ( 55.915  )    GET 055:54:58   ( 55.916 )          SITE
      ——— LIFE SUPPORT ———                  ——— PRIMARY COOLANT ———
GF3571   LM CABIN P   PSIA              CF0019   ACCUM QTY   PCT    34.4
CF0001   CABIN P      PSIA      5.1     CF0016   PUMP P      PSID   46.7
CF0012   SUIT P       PSIA      4.1*    SF0260   RAD IN  T   °F     73.8
CF0003   SUIT ΔP      IN H2O   -1.64
CF0015   COMP ΔP      P PSID    0.30
CF0006   SURGE P      P PSIA    891     CF0020   RAD OUT T   °F       35
         SURGE QTY    LB        3.67    CF1081   EVAP IN T   °F     45.9
  O2     TK 1 CAP ΔP PSID      -109*    CF0017   STEAM T     °F       64
  O2     TK 2 CAP ΔP PSID      -872*    CF0034   STEAM P     PSIA  0.161
                                        CF0018   EVAP OUT T  °F     44.2
CF0036   O2 MAN P     PSIA      105
CF0035   O2 FLOW      LB/HR     0.181
                                        SF0266   RAD VLV 1 / 2       ONE
CF0008   SUIT T       °F        50.8    CF0157   GLY FLO   LB / HR 214.6
CF0002   CABIN T      °F          65           ——— SECONDARY COOLANT ———
CF0005   C02 PP       MMHG       1.5    CF0072   ACCUM QTY   PCT    36.8
      ——— H2O ———                       CF0070   PUMP P      PSID    9.1
CF0009   WASTE        PCT       24.8    SF0262   RAD IN  T   °F     76.5
         WASTE        LB        13.9    SF0263   RAD OUT T   °F     46.2
CF0010   POTABLE      PCT      104.1    CF0073   STEAM P     PSIA  .2460
         POTABLE      LB        37.5    CF0071   EVAP OUT T  °F     66.3
CF0460   URINE NOZ T  °F          72    CF0120   H20-RES     PSIA   25.8
CF0461   H20 NOZ T    °F          72    TOTAL    FC CUR      AMPS  81.45
      ——— CRYO SUPPLY ———      02-1 ——— 02-2 ——— H2-1 ——— H2-2 ———
SC0037-38-39-40  P          PSIA     782 *    19 *    224.2 *   233.6 *
SC0032-33-30-31  QTY        PCT    78.04    47.04     73.64     74.03
SC0041-42-43-44  T          °F     -190       84      -417      -416
                 QTY        LBS   252.4    260.0     20.72     20.83
```

Leaking O_2 Tank 1 has now dropped to 377 psia

```
LM1839                        CSM ECS-CRYO TAB                          0613
CTE 055:58:24 ( 55.93  )     GET 055:58:26   ( 55.974 )          SITE
      ——— LIFE SUPPORT ———                  ——— PRIMARY COOLANT ———
GF3571   LM CABIN P   PSIA              CF0019   ACCUM QTY   PCT    35.6
CF0001   CABIN P      PSIA      5.1     CF0016   PUMP P      PSID   46.9
CF0012   SUIT P       PSIA      4.1*    SF0260   RAD IN  T   °F     56.4
CF0003   SUIT ΔP      IN H2O   -1.60
CF0015   COMP ΔP      P PSID    0.30
CF0006   SURGE P      P PSIA    891     CF0020   RAD OUT T   °F       34
         SURGE QTY    LB        3.67    CF1081   EVAP IN T   °F     44.6
  O2     TK 1 CAP ΔP PSID      -514*    CF0017   STEAM T     °F     64.4
  O2     TK 2 CAP ΔP PSID      -872*    CF0034   STEAM P     PSIA   .162
                                        CF0018   EVAP OUT T  °F     44.0
CF0036   O2 MAN P     PSIA      105
CF0035   O2 FLOW      LB/HR     0.181
                                        SF0266   RAD VLV 1 / 2       ONE
CF0008   SUIT T       °F        50.5    CF0157   GLY FLO   LB / HR 211.9
CF0002   CABIN T      °F          65           ——— SECONDARY COOLANT ———
CF0005   C02 PP       MMHG       1.5    CF0072   ACCUM QTY   PCT    36.8
      ——— H2O ———                       CF0070   PUMP P      PSID    9.3
CF0009   WASTE        PCT       25.6    SF0262   RAD IN  T   °F     76.8
         WASTE        LB        14.3    SF0263   RAD OUT T   °F     47.4
CF0010   POTABLE      PCT      104.1    CF0073   STEAM P     PSIA  .2460
         POTABLE      LB        37.5    CF0071   EVAP OUT T  °F     65.7
CF0460   URINE NOZ T  °F          73    CF0120   H20-RES     PSIA   25.8
CF0461   H20 NOZ T    °F          76    TOTAL    FC CUR      AMPS  61.29
      ——— CRYO SUPPLY ———      02-1 ——— 02-2 ——— H2-1 ——— H2-2 ———
SC0037-38-39-40  P          PSIA     377 *    19 *    228.7 *   236.6 *
SC0032-33-30-31  QTY        PCT    74.81 *  -103 *    74.05    -1.24 *
SC0041-42-43-44  T          °F     -195     -329 *    -417      -427
                 QTY        LBS   241.9    260.0     20.84     -0.35
```

— Appendix D —
Apollo 13: Coincidences of the Number 13

There were multiple occurrences of the number 13 that had become associated with the mission.

So, "coincidence" some people say, but here are some of them:
o *Apollo 13* lifted off pad 39 (a multiple of 3 x 13),
o 13 minutes after the 13th hour, CST on April 11, 1970.
o If you add up all the numerals in the launch date of 4/11/70 they equal 13.
o The accident took place on April 13, 1970.
o The Retrofire Officer's console in the MOCR was numbered 13
o There are 13 letters in "German Measles," the illness that grounded Ken Mattingly, the original Command Module Pilot.
o There are 13 letters in the names "James, Fred, and Jack" as in James Lovell, Fred Haise, and Jack (or John) Swigert, the crew of *Apollo 13*.
o Horoscope for *Aquarius* (LM) from the *Houston Post* – April 13, 1970: *"Do surprises turn you on? Then this is the day for the unexpected."*

And, of course, it's been tried with the words "Apollo Thirteen," but they added up to 14, not 13. I guess that blows it. Are we sure Apollo has two els? And if Ken Mattingly had flown instead of Jack, the names would have added up to twelve. And that means ... ?

— Appendix E —
Those Who Have Passed: In Memoriam

Will Fenner:	GUIDO
Bobby Spencer:	RETRO
Larry Canin:	GNC
Mel Brooks:	Agena
Jim I'Anson:	RETRO
Herb Harman:	CSM SCS
Harry Smith:	Gemini / Agena / LM
Dick Palmantier:	CSM
Dick Brown:	CSM EPS
Bob Britton:	O&P
Julius Conditt:	NC
Cliff Charlesworth:	FD
Perry Ealick:	INCO
Ed Marzano:	AF LM
Larry Armstrong:	O&P
Chris Critzos:	Flight Operations Admin. Assistant
Stu Davis:	Mercury / Gemini systems – MCC M&O
Lyn Dunseith:	MCC Mission Support
Myles Franklin:	Remote Site Systems / MCC Experiment Systems
Scott Hamner:	BOOSTER
Frank Janes:	Recovery
John Mayer:	Mission Planning & Analysis
Duane Mosel:	MCC Flight Procedures
Jim Moser:	Mercury / Gemini Remote Site Systems
Gene Muse:	Gemini Remote Site Systems
Gene Duret:	Mercury Operations Planning
Will Presley:	GUIDO
Gene Chmielewski:	Training / Skylab
Sig Sjoberg:	FOD Mgr.
Howard Tindall:	Flight Design
Ted White:	Remote Site & MCC Systems
Bill Garvin:	Remote site CapCom / Systems
Quarance Patin:	LM Systems
Carl Huss:	Flight Design
Dick Thorson:	LM Control
Bill Yeaky:	Flight Training
John Hatcher:	M&O
George Bliss:	CSM ECS
Tal Sulmeisters:	CSM
Larry Wafford:	Remote Site systems
Stu Present:	pilot, AFD
Don Bourque:	LM systems
Don Puddy:	Flight Director
Pete Frank:	Flight Director
Chuck Filley:	Recovery
Bob Myers:	Training
Merril Lowe:	Requirements
Frank Digenova:	CSM G&N
Jim Nelson:	LM ECS
Chuck Stough:	FAO
John Harpold:	Mission Planning & Analysis

Appendix F
Acronyms and Abbreviations

ADD:	Attention Deficit Disorder
A/G:	Air to Ground
AGS:	Abort Guidance System
AFD:	Assistant Flight Director
ALSEP:	Apollo Lunar Scientific Experiment Package
AOS:	Acquisition of Signal
ARIA:	Apollo Range Instrumentation Aircraft
AS:	Apollo Saturn
ASPO:	Apollo Spacecraft Program Office
ASTP:	Apollo-Soyuz Test Project
ATM:	Apollo Telescope Mount
AUX:	Auxiliary
AWOC:	Army Weather Observers Corps
BBQ:	Barbecue
BDA:	Bermuda (tracking site)
BPC:	Boost Protective Cover
BSE:	BOOSTER (Booster Systems Engineer)
CapCom:	Capsule Communicator
CBRM:	Charger Battery Regulator Module
CCATS:	Communications, Command, And Telemetry System
CDDT:	Count Down Demonstration Test
CDR:	Commander
CM:	Command Module
CMG:	Control moment Gyro
CMP:	Command Module Pilot (sometimes pronounced Simp)
CO:	Commanding Officer:
CONTROL:	Guidance, Control, and Propulsion Officer for the LM
Cryo-Stir:	Cryogenic Gas Storage System Mixing Procedure
CSM:	Command and Service Module
CRT:	Cathode Ray Tube
DFM:	Dumb Fudge Medal
DoD:	Department of Defense
ECS:	Environmental Control System
EDS:	Emergency Detection System
EECOM:	Electrical, Environmental, Sequential Systems Engineer (CSM)
EGIL:	Electrical, General Instrumentation & Life Support
EI:	Entry Interface
ELSC:	Earth Landing Sequence Controller
EMS:	Entry Monitor System
EMU:	Extravehicular Mobility Unit (Space Suit)
EPS:	Electrical Power System
EVA:	Extra Vehicular Activity
FAO:	Flight Activities Officer
FC:	Fuel Cell
FCOB:	Flight Control Operations Branch
FCR:	Flight Control Room
FD:	Flight Director
FDO:	Flight Dynamics Officer (FIDO)
FOD:	Flight Operations Division / Director
GAEC:	Grumman Aircraft Engineering Corp.
GET:	Ground Elapsed Time
GNC:	Guidance, Navigation, and Control officer
GSE:	Ground Support Equipment
GUIDO:	Guidance Officer
H_2:	Hydrogen
HOSC:	Huntsville Operations Support Center
HGA:	High Gain Antenna
INCO:	Instrumentation and Communications Systems Officer
ISS:	International Space Station
JEC:	Johnson Engineering Corporation
JFK:	John Fitzgerald Kennedy
JSC:	Johnson Space Center (Formerly MSC)
KP:	Kitchen Police:
KSC:	Kennedy Space Center
LACC:	Los Angeles City (Jr.) College
LES:	Launch Escape System
LGE:	Lunar Geology Engineer

LOI:	Lunar Orbit Insertion	**PI:**	Principal Investigator
LM:	Lunar Module	**PLSS:**	Potable Life Support System
LMP:	Lunar Module Pilot (sometimes pronounced Limp)	**P-Tube:**	Pneumatic Tube
LOS:	Loss Of Signal / Line of Sight	**PSIA:**	Pounds per square inch absolute
LRD:	Landing and Recovery Division	**PTC:**	Passive Thermal Control
		Q-Ball:	Trajectory Deviation Indicator (Tip of LES)
LSPO:	Lunar Surface Planning Officer	**RA:**	Regular Army
MAP:	Message Acceptance Pulse	**R&D:**	Research And Development
MC&W:	Master Caution and Warning	**RCS:**	Reaction Control System
MCC-H:	Mission Control Center – Houston	**RETRO:**	Retrofire Officer
		ROCR:	Recovery Operations Control Room
MIT:	Massachusetts Institute of Technology	**RTCC:**	Real Time Computer Complex
MCP:	Mission Control Programmer	**S&ID:**	Space and Information Systems Division (NAA)
MER:	Mission Evaluation Room		
MESC:	Mission Events Sequence Controller	**SCA:**	Simulation Control Area
		SCE:	Signal Conditioning Equipment
MET:	Mission Elapsed Time	**SCS:**	Stabilization and Control System
MFSC:	Marshall Space Flight Center		
MMU:	Manned Maneuvering Unit	**SCSC:**	Space City Ski Club
MOCR:	Mission Operations Control Room	**SECS:**	Sequential Events Control System
MOW:	Mission Operations Wing (Building 30)	**SIMSUP:**	Simulation Supervisor
		SLA:	Spacecraft / Lunar Module Adapter
MP:	Military Police:	**SM:**	Service Module
MPAD:	Mission Planning and Analysis Division	**SPAN:**	Spacecraft Planning and Analysis room
MPSR:	Multi-Purpose Support Room	**SPS:**	Service Propulsion System
MSC:	Manned Spacecraft Center (Now JSC)	**SSR:**	Staff Support Rooms
		STG:	Space Task Group
MSFC:	Marshall Space Flight Center	**TCE:**	Fuel Cell Condenser Exhaust Temperature
NAA:	North American Aviation		
NAR:	North American Rockwell	**TD&E:**	Transportation, Docking and Extraction
NASA:	National Aeronautics & Space Administration		
NBA:	National Basketball Association	**TEC:**	Trans-Earth Coast
		TEI:	Trans-Earth Injection
NC:	Network Controller	**TELCOM:**	Thermal, Electrical, and Communications
NCO:	Non-Commissioned Officers		
NETWORK:	Network Controller	**TELMU:**	Telemetry, Electrical, and EMU
NR:	North American Rockwell		
O&P:	Operations and Procedures	**TLC:**	Trans-Lunar Coast
OCS:	Officer Candidate School	**TLI:**	Trans-Lunar Injection
OCT:	Officer Candidate Test	**TLM or TM:**	Telemetry
OD:	Order of the Day, Olive Drab	**TNT:**	Trinitrotoluene
OJT:	On the Job Training	**VAB:**	Vehicle (formerly Vertical) Assembly Building
OSH:	Off Scale High		
O₂:	Oxygen	**VDC:**	Volts Direct Current
OWS:	Orbiting Work Shop	**VOX:**	Voice-Operated Transmission
PGNS:	Primary Guidance and Navigation System (LM)	**YTS:**	Yuma Test Station

CDROM

The attached CDROM features:
More than 3 hours of NASA tape audio from the Mission Control EECOM loop during the *Apollo 13* and *Apollo 15* missions.
A Quicktime panorama of the NASA Mission Operation Control Room
A 45 minute MPEG video presentation about the *Apollo 13* explosion by Sy Liebergot
A picture gallery of Apollo-era Mission Controllers at work and play.
"I Don't Understand," roast song.
The disc is rated for Windows 95 and higher. An adequate web browser is also necessary to play the disc.

About the author

Sy Liebergot

Seymour A. (Sy) Liebergot was a NASA flight controller during the exciting Apollo and post-Apollo years. His many experiences include a critical role in the safe recovery of the crew of the ill-fated *Apollo 13* mission. Sy still works in the U.S. space program as an ISS Senior Configuration Management Analyst for Spacehab Incorporated at Johnson Space Center.

About the contributing editor

David M. Harland is the author of a number of well received space science and exploration books including, *Mission to Saturn – Cassini and the Huygens Probe*, *The Earth In Context – A Guide to the Solar System*, and *Jupiter Odyssey – The story of NASA's Galileo Mission*, among others. Well recognized and respected in the space industry, David Harland lives in Glasgow, Scotland.